ENTER THE DRONES

The **FAA** and **UAVs** in **America**

BILL CAREY

Schiffer Publishing Ltd

Other Schiffer Books on Related Subjects:
Unmanned Air Vehicles: An Illustrated Study of UAVs, Bill Holder, ISBN: 978-0-7643-1500-8

Library of Congress Control Number: 2016932849

Designed by Molly Shields
Cover design by Justin Watkinson
Type set in Impact/Times New Roman

ISBN: 978-0-7643-5077-1
Printed in China

Published by Schiffer Publishing, Ltd.
4880 Lower Valley Road
Atglen, PA 19310
Phone: (610) 593-1777; Fax: (610) 593-2002
E-mail: Info@schifferbooks.com
Web: www.schifferbooks.com

DEDICATION

For my parents, William and Patricia, and my wife, Elena.

CONTENTS

ACKNOWLEDGMENTS

My sister Mary—holder of a PhD in American literature and a better writer than I—provided invaluable advice and support during the many months turned into years it took to complete this history. Formerly a reporter herself, she has instructed the next generation of journalists as an adjunct professor at the University of Massachusetts-Amherst. My lifelong friend from Pittsfield, Massachusetts, Steve Harvey—now a successful Philadelphia attorney and ardent environmentalist—provided needed advice on the chapter devoted to privacy and free speech. My good friend and former colleague Ron Laurenzo—commercial pilot, journalist, accomplished pianist, and master of languages—vetted aeronautical information. I would like to thank all of the people I interviewed for taking the time to answer my questions and sharing their wisdom. The Federal Aviation Administration—especially public affairs specialist Les Dorr—responded in timely fashion to all of my inquiries. The US Army and Air Force were particularly helpful in making experts available to speak with me, as was MITRE Corporation. Photographer Bill German supplied a beautiful photograph of the *Westward Wind* and also provided me with important insight into the maritime industry. The media specialists of Insitu answered my many requests and supplied the photograph of the ScanEagle crew from the ship's historic mission. I thank my employer, *Aviation International News*, for allowing me to write this book on the margins of my full-time reporting job. Many a weekend hour was spent in the library at American University in Washington, DC. Randy Padfield, a veteran Air Force and commercial helicopter pilot, author, and journalist who retired as AIN's chief operating officer in 2014, helped me submit the proposal for this book; Helicopter Association International named him the winner of its 2016 excellence in communications award, and rightfully so. Thanks to Schiffer Publishing for accepting my proposal to write a history on civil and commercial drones, and to editor Ian Robertson for his patience in accommodating my many questions and late changes. This book is dedicated to my late parents, who made everything possible for me, and to my wife, Elena, who has stayed with me throughout. No elaboration is needed.

INTRODUCTION

This is not a military history of drones. Other writers have covered the military legacy of unmanned aircraft—drones, UAVs, what have you—far better than I could have. Rather, this book is an attempt to describe the emergence of drones as a civil and commercial phenomenon in the United States. That legacy dates to at least the early 1990s when considering civil government applications, but this book focuses on the formative years 2012 to 2015, the period following passage of the landmark FAA Modernization and Reform Act of 2012. Spurred by the will of Congress, the Federal Aviation Administration moved from its dated prohibition of money-making flights to approving thousands of commercial exemptions to operate drones. Authorizations were no longer granted by exception but as a matter of routine. A risk-averse federal bureaucracy structured along lines of business that one knowledgeable person described as "tribal," pressured from without by other branches of government and aviation's many constituencies, had turned on a dime. The FAA said all along it would introduce unmanned aircraft into the National Airspace System safely and incrementally, starting with the smaller machines. Ultimately, though, it fell short of at least the expectation created by the 2012 congressional mandate to "provide for the safe integration of civil unmanned aircraft systems into the national airspace system as soon as practicable, but not later than September 30, 2015." An FAA final regulation that will shape the entry of small drones into the airspace in a uniform way—several drafts and years in the making—was finally expected around the time of this book's release.

My earliest recollection of the topic dates to 2007. I was a panel moderator for an airline-oriented conference in Amsterdam, and two of the panelists who spoke were "prominent advocates of mainstreaming unmanned aircraft in civil airspace," I later wrote. The duo of American John S. "Johnny" Walker and Dutchman Peter van Blyenburgh "challenged, even cajoled the audience to get involved" and made for "an entertaining and informative presentation." The airline crowd seemed largely uninterested and my impression was the topic was a bad fit for the conference, but I was intrigued and continued to follow unmanned aircraft. I stayed in touch with Walker. A former FAA executive, he was then co-chairman of a committee assembled by standards organization RTCA, née the Radio Technical Commission for Aeronautics, to address some of the biggest technical barriers preventing unmanned aircraft from entering the airspace—none more challenging than their need to "sense and avoid" other aircraft. I attended most annual gatherings of the Association of Unmanned Vehicle Systems International, the industry's most prominent trade organization. Mainly military focused, AUVSI's Unmanned Systems Conference was an outlier among aviation tradeshows.

When drones blundered onto the domestic stage after 2010, major media outlets tended to emphasize their potential for intrusiveness, and only later the safety hazard they presented. *Time* magazine's cover from February 11, 2013, left a lasting impression; it depicted an armed MQ-1 Predator hanging menacingly above a suburban home. "They are America's global fighting machines," the cover copy instructed. "What happens when they're unleashed at home?" The thought that recurred when I read or watched such coverage was *it's more complicated than you think*. The inside-baseball nature of trade press reporting taught me that flying unmanned aircraft in civilian airspace was complex from a regulatory perspective, so I set out to explain the complexities.

My preconceived notions about writing this book evolved with the telling. This may not be a military history of drones, but the military's contribution past and present cannot be diminished. The US Department of Defense maintains the world's largest fleet of unmanned aircraft and continues to drive new applications and capabilities of the technology. Peace activists did not figure in my initial thinking, but it was Code Pink, among other groups, that helped frame the way Americans perceived drones; it brought the dirty wars home in a vivid way. Hollywood also deserves credit and, of course, the media. I didn't plan a chapter on public UAVs, but came to appreciate that agencies such as the Department of Energy were civilian pioneers of unmanned aircraft; they paved the regulatory road on which private entities would follow. I wanted to avoid writing about hobby drones, but they demanded attention—not just my attention, but the FAA's. Irresponsible flights by some hobbyists near airports, over crowds, and within emergencies invited more regulation for all. Early on, I entertained the thought of chronicling each rogue drone event that made the newspapers or the evening news, but it became apparent this was a fool's errand.

I wanted to capture as many voices as possible, but no doubt there will be deserving individuals, institutions, and companies that I have missed. There were some people I would have liked to interview who either declined or did not respond to my requests, attesting to the secrecy that still attends drones or maybe a general mistrust of journalists, so they have left me to tell the story.

CHAPTER 1

FIRST COMMERCIAL OPERATION

On September 12, 2013, a modified crab catcher-processor ship navigating the Chukchi Sea north of Alaska helped write aviation history. From the forward deck of the *Westward Wind*, a four-man crew launched a small, remote-controlled airplane into the Arctic wind. The Insitu ScanEagle weighed thirty-five pounds. It had a five-foot-long, cigar-shaped body with a pusher propeller in the rear that drove it forward, a needle of a nose, and a glass bulb for a chin. Its wings were swept back, measuring ten feet across to where they turned upward as winglets. The National Weather Service forecast for the Chukchi Sea coast that day called for periods of rain developing in the morning, areas of fog, and temperatures in the mid-to-upper 40s. The ScanEagle took wing when the ship was about 120 miles off the village of Wainwright, on Alaska's North Slope; it flew for thirty-six minutes. On a final pass of the *Westward Wind* one of its wingtips snagged a rope hanging from a boom on the ship and the little airplane dangled in the air like a fly caught in a spider's web.

Launching an airplane from a ship off the coast of Alaska was by definition demanding, but that was not what made this flight noteworthy, nor was this the ScanEagle's first flight in austere conditions—not by a long shot. The *Westward Wind's* mission was more precisely a milestone in US aviation regulatory history: it was the first Federal Aviation Administration approved flight of an unmanned aerial vehicle for a commercial purpose. The customer was Houston-based oil exploration and production company ConocoPhillips; the service provider was Insitu. A year later, it would be Hollywood that grabbed the spotlight for putting robotic aircraft to work. The FAA's next major round of permits went to several aerial photo and video production companies that used miniature helicopters fitted with cameras to make movies. Until then, movies including the James Bond thriller *Skyfall* that used UAVs in production had to be filmed overseas, said Christopher Dodd, former US senator from Connecticut. Dodd now headed the Motion Picture Association of America, which supported the video companies in their applications to the FAA. "Today's decision will create a climate where more production is done at home," he declared when the permits were announced. "It's also important to note that it will encourage creative, cutting-edge technology as well that will allow us to make scenes and pictures even more imaginative and creative than they have been in the past." Hollywood's prominent role in introducing UAVs into America's airspace was fitting for a technology that was still futuristic and somewhat fictional in the popular imagination.

Insitu's original product was the Aerosonde, a robotic airplane that it designed in the mid-1990s with funding from the US Office of Naval Research and co-developed with an Australian company (Environmental Systems and Services of Melbourne) and the Australian Bureau of Meteorology to take aerial weather measurements. On August 20, 1998, a team including Insitu co-founder Tad McGeer and University of Washington researchers launched an Aerosonde named *Laima* (for the Latvian deity of good fortune) from the roof of a rental car on Bell Island, Newfoundland. Twenty-six hours, forty-five minutes later, the twenty-eight-pound machine landed in tall grass on South Uist, in Scotland's Outer Hebrides islands—the first Atlantic crossing by an unmanned aerial vehicle. Insitu next developed the SeaScan airplane with the commercial fishing industry in mind, serving as an aerial platform for spotting schools of tuna. With it the company built other pieces of a turnkey unmanned aircraft system: a pneumatic catapult launcher, a "SkyHook" apparatus to retrieve the airplane in flight, and a gyro-stabilized video system to scan the ocean's surface. The SeaScan was not a commercial success, but the technology interested aerospace giant Boeing. In February 2002, the companies entered into an agreement to co-develop a prototype based on the SeaScan called ScanEagle, which contained a gimbaled sensor turret with day and night cameras in its glass chin. A year later, the ScanEagle and other unmanned aircraft models showed their value in providing "intelligence, surveillance, and reconnaissance" by sharing data and video within a network of forces during a US Navy exercise in the Bahamas.

The US Marine Corps First Marine Expeditionary Force brought the ScanEagle and its Boeing operators to Iraq, where in late 2004, the birdlike aircraft participated in the second battle to wrest control of the city of Fallujah from insurgents. Capable of streaming video to ground stations and remote video terminals, the ScanEagle provided commanders with a forward look at enemy positions and threats, by all accounts

The Insitu ScanEagle went to war with the US Marines in Iraq in 2004. Oil company ConocoPhillips and Insitu launched one from a converted fishing vessel in the Chukchi Sea on September 12, 2013, marking the first time the FAA approved a commercial unmanned aircraft flight. *Courtesy of Insitu.*

preventing casualties. "It's called ScanEagle, and it has already saved the lives of many Marines," the American Forces Press Service later reported. The ScanEagle "had its baptism of fire during some of the heaviest urban combat Marines have been involved in since Hue City in Vietnam in 1968." It "performed flawlessly, First Marine Expeditionary Force officials said."[2] Over the next several years, the US Navy and Special Operations Command, the Australian Army, Canadian Forces, the UK Royal Navy, and the militaries of Poland, the Netherlands, and Singapore ordered the ScanEagle service Insitu developed. In the meantime, Boeing acquired Insitu in September 2008 for a reported $400 million, adding the weight of an aerospace manufacturer with $61 billion in sales that year to what had been a small start-up company in Bingen, Washington, a village on the Lewis and Clark Highway in the Columbia River Gorge. "This acquisition is part of a larger plan to aggressively grow our presence in the unmanned systems market," declared Chris Chadwick, then president of Boeing's St. Louis-based military aircraft division, to which Insitu was assigned.[3]

An aerial platform with cameras could be useful for more than just tracking enemy combatants. The ScanEagle made inroads among research universities and government agencies testing unmanned aircraft on the US domestic front, with much of the activity centered in and around Alaska. The University of Alaska-Fairbanks Geophysical Institute purchased a ScanEagle in 2006 to study the airplane's applications for mapping wildfires, patrolling oil pipeline, and monitoring movements of large animals at its Poker Flat Research Range. "They have a large rocket range that they were using to do sounding rocket research, and they were going to leverage that along with ScanEagle to get the ball rolling for utilizing unmanned systems in other than defense-related activities," recalled Insitu's Paul McDuffee, who led the company's effort to branch into civil and commercial markets. "That was the beginning." The University of North Dakota sponsored a series of ScanEagle flights that Evergreen International Aviation flew for Shell Oil Co. in the Arctic Ocean in 2007 and 2008.[4]

After a hiatus of more than a decade, during which it concentrated its efforts in the Gulf of Mexico, Shell had "re-entered Alaska" in 2005 by acquiring hundreds of exploration leases in the Beaufort Sea and three years later in the Chukchi Sea.[5] The test flights Evergreen flew on its behalf in the late summer and fall of 2008 were concentrated in Harrison Bay—part of the Beaufort Sea—and involved five ScanEagles. The objective was to measure the ScanEagle's performance against manned Twin Otter turboprops with trained observers on board spotting marine mammals in ocean tracts targeted for "seismic" and other exploration activities. Seismic exploration involved using arrays of airguns towed behind ships to fire shock waves of compressed air into subsurface rock layers and then measuring the returns with acoustic receivers to determine the presence of oil and gas deposits; this had the potential to injure or kill nearby animals. Evergreen personnel controlled the ScanEagle from the research vessel *Norseman II*. The aircraft's task was to detect twelve inflatable kayaks the researchers peppered off the coast to simulate whales and to record any "opportunistic" sightings of marine mammals. "Information from these flights will be used to verify whether unmanned surveys are to be an accepted method of meeting and monitoring mitigation requirements issued by the National Marine Fisheries Service and US Fish and Wildlife Service," the *Alaska Journal of Commerce* reported.[6] Project participants claimed the eleven test flights were the first non-military use of an unmanned aircraft in the US outside of special or restricted airspace. The project found unmanned aircraft "proved to be functional in the Arctic."[7]

The National Oceanic and Atmospheric Administration and the University of Alaska-Fairbanks conducted ScanEagle flights from the fishery survey vessel *Oscar Dyson* over the Puget Sound, in Washington state, in October 2008, and from the research vessel *McArthur II* in the Bering Sea in June 2009. The latter mission was to evaluate the aircraft's use in determining the "abundance and distribution of ice seals" and was the first civil-approved "beyond line of sight" operation—that is, one in which the pilot can no longer see the aircraft—according to the university. The Coast Guard demonstrated the ScanEagle from the cutter *Stratton* off the California coast in August 2012. A ScanEagle the service flew from the cutter *Bertholf* scored another first when it helped stop a drug-running operation off the west coast of Central America in May 2013, months before the *Westward Wind's* historic mission.

The Regulatory Environment

Until that mission, the FAA—as overseer of the nation's airspace system—allowed only two means of flying unmanned aircraft in civil, unrestricted airspace, or airspace that was available to other users and not fenced off for military or testing purposes. Military services, governmental agencies, and public universities needed a certificate of authorization or waiver commonly called a COA, which was essentially operational approval the FAA granted to use a defined block of airspace for a specific activity. They needed to prove the aircraft was "airworthy," or in a safe condition to fly, under either FAA or Department of Defense certification criteria. Private-sector companies needed something different: a special airworthiness certificate in the experimental category. The FAA issued these for developmental flying machines that lacked a design approval known as a type certificate. With an experimental "ticket" in hand, the operator could fly an unmanned aircraft for research and development purposes, crew training, or sales demonstrations, but not for profit. In 2007, as the number of operations multiplied, the FAA issued a policy notice that explicitly forbade flying unmanned aircraft for "business purposes." It convened a "small UAS" advisory committee in 2008, to help it draft rules that would allow the operators of small unmanned aircraft systems—defined as those weighing fifty-five pounds or less—to gain access to the airspace. The committee delivered its recommendations in April 2009. Several drafts of the so-called small UAS rule circulated within the FAA and its parent organization, the Department of Transportation, but years later, the agency had not released the rule for comment.

The FAA's two-track authorization process, while accommodating public agencies and universities, slowed and frustrated a simmering commercial industry, causing concern that the United States was ceding its technological edge to countries with more flexible regulations, and Congress reacted to the mounting pressure in 2012. After passing twenty-three "stopgap" spending bills that only temporarily extended the FAA's operating authority, it finally approved long-term legislation—the FAA Modernization and Reform Act of 2012—that contained important provisions for the advancement of commercial UAVs. An overarching provision required the FAA to provide for "safe integration" of unmanned aircraft into the national airspace system by September 30, 2015. Another required it to "designate permanent areas in the Arctic where small unmanned aircraft may operate 24 hours per day for research and commercial purposes." The act defined "Arctic" as the US zone of the Chukchi Sea, the Beaufort Sea, and the Bering Sea north of the Aleutian island chain—remote expanses where operators could fly robotic aircraft beyond line of sight relatively safely. Although off the US coast these were international waters, the International Civil Aviation Organization assigned responsibility for the airspace north and west of Alaska to the FAA, which designated it the Anchorage flight information region. The Arctic provision was one path the FAA could pursue to satisfy the congressional mandate.

In March 2012, a month after President Barack Obama signed the "FMRA" legislation, the FAA combined offices from the different organizations within the agency that dealt with unmanned aircraft into a single Unmanned Aircraft Systems Integration Office and appointed agency veteran Jim Williams to run it. "When I first came on board," Williams recalled in an interview, "I started asking the troops here: how can we do early approvals? I went up to Alaska just to kind of get to know people in the summer or early fall of 2012 because we had legislation saying 'you've got to open up the waters in Alaska.' I'm like, God, how are we going to do this? We need some sort of demonstration program that will help us figure this out."

By Williams's reckoning, there were three things operators needed to fly an unmanned aircraft for commercial purposes: they needed a qualified pilot at the controls, a requirement that could be fulfilled by a person with a pilot's certificate and the aeronautical knowledge that conveyed, who was also trained on the specific unmanned model; and they needed an operational approval spelling out the flight parameters, time, notification requirements, and safety precautions of an activity, something the FAA already granted to public agencies through the COA process. The third leg of the stool was problematic: they needed an aircraft with a type certificate proving it was "properly designed and manufactured" to manned aircraft standards, or an airworthiness certificate proving it conformed to its type design and was in a safe condition to fly—approvals that unmanned machines built for military or research purposes did not have. But now that the FAA had the will of Congress to consider it found a way to accommodate them. Whereas the experimental category of certification rules allowed an aircraft with no type certificate to conduct a limited set of operations, the "restricted" category allowed prior military acceptance of an aircraft to serve as the basis for a type certificate, supporting a wider set of operations. The aircraft was to be "of a type that has been manufactured in accordance with the requirements of and accepted for use by an Armed Force of the United States and has been later modified for a special purpose," stated federal aviation regulations Part 21.25. Allowable special purposes under the restricted category were agricultural spraying, forest and wildlife conservation, aerial surveying, pipe and powerline patrol, cloud seeding, aerial advertising, and "any other operation specified by the FAA."

Since oil companies "were a big push" behind the FMRA legislation, Williams said he spoke with Shell, BP, and ConocoPhillips about their respective views on using unmanned aircraft. He also consulted with Insitu's Paul McDuffee, who had served on the small UAS advisory committee and another "ARC" (Aviation Rulemaking Committee) the FAA had convened in 2011 to help it develop standards for a broader range of unmanned aircraft systems. "Insitu and ConocoPhillips had been considering options for using unmanned aircraft for several years," McDuffee recalled. "We were trying to work within the regulations as they existed at the time, i.e., finding out a way to fly these airplanes as a state or government-operated aircraft, under a COA, and yet provide information that would be important and valuable to a commercial customer," he said. "It just got more problematic as we went down the road—finding the right government sponsor for the operation, who owns the data, and all of that. We weren't able to put anything together of substance." What finally "moved the needle," he said, was the passage of the FMRA, with its provisions to speed the introduction of unmanned aircraft. "We being Insitu had a lot of input in (the legislation), including the Arctic language that's in the FMRA," McDuffee explained. "We specifically wanted this added, that the FAA would establish permanent UAS more or less free-fly zones in the Bering, Chukchi, and Beaufort seas off the west and northwest coast of Alaska for small UAS operations below 2,000 feet above the surface. Surprisingly, that language got included in the FMRA almost verbatim from the way we proposed it. It was as big a surprise to us as it was to anybody else, but it got in there."

McDuffee said Williams deserved credit for "the brainstorm about looking at ways to adapt" Part 21.25 to certify an unmanned aircraft.

"It had never been done with a UAS before," he noted. McDuffee went on to say:

> We hooked up with the (FAA's) Los Angeles certification office upon the recommendation of some good friends within the FAA and the rest is history. We spent a good year knocking out a lot of the regulatory and technical hurdles and the complete review of the Part 21.25 requirements for taking that prior military acceptance and applying it to an unmanned system, looking at all the operational regulations for flying in the national airspace in that particular region, running the whole process through the safety-risk management process the FAA mandates—all that was completed. There were no regulations basically cut out of the loop or just ignored. We had to address each and every one of them to get the type certificate and also the operational approval to begin flights up there.

With its new approach to certifying unmanned aircraft, the FAA and ConocoPhillips entered into an "other transaction authority" agreement that authorized the oil company to use the ScanEagle—flown by Insitu under restricted category type certification—to support its oil and gas exploration activities in the Chukchi Sea. In exchange, ConocoPhillips agreed to share flight data with the FAA. Williams said he invited other companies to participate so the agency's creative certification approach was not exclusive to ConocoPhillips and Insitu. California-based manufacturer AeroVironment, which had supplied thousands of hand-launched Wasp, Raven, and Puma airplanes to the US military since the mid-2000s, was also interested.

"We had a lot of hurdles to get over, and that was the whole point," Williams said. "By going through the process of getting the aircraft certified and getting the operation approved, we were able to ferret out what are the problems, what are the incompatibilities between the current manned aircraft processes and unmanned aircraft?" Kevin Hull, the FAA's Los Angeles aircraft certification office manager, "really latched onto this as a great program," Williams added. "He wanted to make it work, so he threw all kinds of resources at it, some of his best people, and we were able to get it done in record time."

In late July 2013, under the heading "One Giant Leap for Unmanned-kind," the FAA announced it had awarded restricted-category type certificates on July 19 to Insitu for the ScanEagle and to AeroVironment for the Puma, allowing them to fly those models commercially for aerial surveying, one of the restricted category's special purposes. The press release did not name the end customers, although *Aviation International News*, reporting remarks that Williams gave to the Air Line Pilots

Association, identified ConocoPhillips as one customer within days of the certificates being granted.[8] According to the FAA's official announcement, "a major energy company" planned to use the ScanEagle to survey ice floes and migrating whales off the Alaskan coast. The expectation for the Puma was that it would be used to support oil spill and wildlife monitoring in the Beaufort Sea. Only later would BP Exploration Alaska be revealed as AeroVironment's customer.

Watching Whales and Ice Floes

The Chukchi Sea, named after the indigenous Chukchi people of Siberia, is a funnel of the Arctic Ocean that separates the Russian Far East from Alaska. It is demarcated from the Beaufort Sea to the east by the coastal outcrop of Point Barrow, Alaska, the northernmost point of the United States. The seas are contiguous, but their underwater terrain differs. Much of the Chukchi Sea covers continental shelf and is less than 160 feet deep. The Beaufort Sea is mostly more than 3,280 feet deep; it has a narrow band of continental shelf along the Alaskan and Canadian coasts that slopes steeply to the depths of the Canada basin. The Pacific Ocean nourishes the Chukchi Sea through the Bering Strait and brings warm water in the summer, contributing to an earlier ice melt. Sea ice covers the Chukchi Sea for some eight months a year—encroaching around November and receding in June—whereas the Beaufort Sea has more of a perennial ice cover. Influenced by surface winds and the configuration of the coast, the seasonal ice of the Chukchi Sea is "more mobile and changeable" than the ice that covers the Beaufort.[9]

The so-called "marginal," partially contained seas of the Arctic were the habitat for numerous species of marine mammals, including bowhead, beluga, and gray whales, harbor porpoises, ringed and bearded seals, Pacific walruses, and polar bears. Some of those creatures, such as bowhead whales, were listed as endangered under the 1973 Endangered Species Act. The two seas were thought to contain an abundance of natural resources; according to one US government estimate, the Chukchi Sea alone held anywhere from 2.32 to 40 billion gallons of recoverable oil, as well as large reserves of natural gas.[10] In February 2008, the Department of the Interior's Minerals Management Service—later renamed the Bureau of Ocean Energy Management—auctioned offshore oil and gas leases in the Chukchi Sea for just the third time—earlier sales took place in 1988 and 1991. The agency described Oil and Gas Lease Sale 193 as the most successful in Alaska's history, demonstrating industry's "heightened interest in this area." Energy companies submitted high bids of $2.6 billion to acquire the rights to tracts within an arc spanning twenty-nine million acres led by Royal Dutch Shell's investment of $2.1 billion. ConocoPhillips spent $506 million to acquire ninety-eight leases in the auction; the Houston company planned to drill one or two exploration wells during the open-water season of 2014 in the Devil's Paw Prospect, centered about 120 miles west of Wainwright. Among the permits it needed was an "incidental harassment authorization" from NOAA's National Marine Fisheries Service and a letter of authorization from Interior's Fish and Wildlife Service, approvals that allowed activities which could result in the unintentional death or injury of marine mammals or disrupt subsistence hunting by coastal communities. To satisfy the requirements of the two authorizations, ConocoPhillips planned to station "protected species observers" who were also native Inupiat speakers on the drill rig and supporting vessels and conduct acoustic monitoring using bottom-mounted recorders to measure the impact of drilling activities.[11]

Despite its auspicious beginning, Lease Sale 193 immediately ran into headwinds. Even before the February 6 auction, a coalition of Alaska native groups and big-name environmental organizations filed suit in federal court in Alaska, "arguing that in making its decision to hold the lease sale, [the government] did not adequately weigh the impact oil and gas activities would have on wildlife like polar bears, or on native villages along Alaska's North Slope."[12] US District Judge Ralph Beistline found the environmental impact analysis the government prepared in advance of the lease sale deficient and suspended drilling activities in the Chukchi Sea in July 2010. The judge lifted the suspension a year later, when the Bureau of Ocean Energy Management presented a "revised draft supplemental environmental impact statement" that addressed his concerns. Shell initiated exploratory drilling in the Chukchi and Beaufort seas in 2012, a season that was interrupted the day after it began in September, when a moving ice pack forced the company to pull the drill bit and anchors on its drill ship *Noble Discoverer* in the Chukchi Sea.[13] The season ended on December 31, when Shell's drill barge *Kulluk*, returning to Seattle from service in the Beaufort Sea, broke free from its tow ship in a heavy storm and ran aground near Kodiak Island.[14]

Shell announced in February 2013 that it would "pause" its exploratory drilling in the Alaskan waters that season and plan for a later resumption of activity. ConocoPhillips followed suit a few months later; citing "the uncertainties of evolving federal regulatory requirements and operational permitting standards," the company said it would suspend its Alaskan offshore drilling plans for 2014. The setbacks continued. In January 2014, the US Court of Appeals for the 9th Circuit found that the former Minerals Management Service, in its environmental impact statement, had arbitrarily assumed only 1 billion barrels of oil was recoverable from the Chukchi Sea. The court remanded the matter back to Beistline, who ordered the government to complete a new environmental impact analysis before any drilling resumed. When the Bureau of Ocean Energy Management released a second-draft supplemental environmental impact statement in October 2014, it had revised the estimate of recoverable oil from the Chukchi to 4.3 billion barrels.

While the industry's drill bits were momentarily silenced, the remote Alaskan waters remained an open canvas for testing unmanned aircraft that could ultimately help the oil giants do business in the Chukchi and Beaufort seas. The federal government required them to monitor the activities of marine mammals in their areas of operation, and Shell's earlier research had shown the ScanEagle was a viable option for that mission. NOAA and the Bureau of Ocean Energy Management annually counted marine mammals in the two seas as part of their Aerial Surveys of Arctic Marine Mammals project. They used DHC-6 Twin Otter and Aero Commander 690A twin-engine turboprops fitted with bubble windows to give observers a fuller look-down view of the water. Flying from Barrow or Deadhorse, the airplanes could stay aloft for about five hours when including needed fuel reserves. A ScanEagle catapult-launched from a moving ship and driven by a small piston engine stayed aloft for twenty or more hours, providing its remote operators a live video feed from the cameras in its nose bay. An unmanned aircraft could also serve in an early-warning role to help detect wind-driven ice floes posing a hazard to drilling installations. Oil companies relied on ice forecasting using radar and satellite imagery, in addition to field observations of the sea ice cover. Their exploration fleets sailed with "ice management" vessels—hardened icebreakers that steered ice floes away from vulnerable drill sites. Shell and ConocoPhillips were "aggressively investigating" the option of using unmanned aircraft to support their exploration activities as early as 2006.

At a National Marine Fisheries Service meeting in October, Michael Faust, ConocoPhillips' offshore exploration manager, explained the US military's use of unmanned aircraft in Iraq was driving the rapid development of the technology. "These things fly at, say, 1,000 feet and you can't see them, you can't hear them," Faust said, according to *Petroleum News*. "I don't think there's any issue at all with these aircraft observing whales."[15]

ConocoPhillips started collecting environmental data in the area of the Devil's Paw Prospect in 2006, and two years later, around the time of Lease Sale 193, expanded the effort into an "ecosystem-based studies program."[16] Olgoonik Fairweather—a joint venture of the native-owned Olgoonik Corporation of Wainwright and Fairweather Science that provided scientific support services to the oil and gas industry—assumed management of the effort in 2010. ConocoPhillips, Shell, and Statoil funded what became the Chukchi Sea Environmental Studies Program. Olgoonik Fairweather's fleet of several research vessels, including the *Westward Wind*, supported the effort. Built in 1978, the *Westward Wind* was formerly owned by Highland Light Seafoods of Seattle and was last licensed as a commercial fishing vessel in 2009, according to the Alaska Department of Fish and Game database. It was a "tough, heavy duty vessel with long voyage capabilities," Olgoonik Fairweather said. Its forward factory was converted into a science laboratory and office space.

The FAA declared the ship's contribution to aviation history eleven days after the fact. "A research vessel plying the icy waters of the Chukchi Sea above the Arctic Circle seems an odd place for a Federal Aviation Administration employee. But that's exactly where Alaskan Region inspector Jay Skaggs was on Sept. 12, 2013," the agency announced in a press release. "His presence aboard the *Westward Wind* helped ensure the first FAA-approved commercial flights by an unmanned aircraft went off safely and without a hitch." The ship carried four ScanEagles, each with its own paperwork—a restricted-category type certificate was for a specific aircraft and serial number, not a series production machine. "Skaggs and the Insitu flight crew led by Jeff Kelly watched as the first commercial ScanEagle zoomed off a catapult and into the rainy Arctic skies," the FAA continued. "After a successful 36-minute flight, the ship's retrieval system captured the UAS and the ground-breaking mission was complete." Kelly was the pilot and team leader; his crew that day included Emory Wanger, responsible for flight deck operations; Thomas Burlace, maintenance; and Mike Sayler, setup operations. Wearing yellow hard hats and orange-and-black coveralls that matched the fluorescent markings on the ScanEagle, they posed for a photo with Charlton Evans, Insitu's commercial UAS program manager, and able-bodied seaman Phil Gray of the *Westward Wind* deck crew.

ConocoPhillips issued its own press release on September 24, 2013, marking the historic feat.* The press release reported that the first flight exercised the aircraft's sensor payload and navigation systems, and while airborne, the ScanEagle sent "real-time" video and telemetry information to the ground control station on the *Westward Wind*. The expectation was the testing would facilitate future flights in the Arctic by streamlining the approval process. "Airborne surveillance is often a component of offshore projects," stated Trond-Erik Johansen, ConocoPhillips Alaska president. "The UAS could be useful in our monitoring and data collection efforts, with the benefit of

* ConocoPhillips did not respond to requests for an interview for this book

The Insitu crew that launched the first FAA-approved commercial flight of an unmanned aircraft system on September 12, 2013, poses with the ScanEagle on the deck of the *Westward Wind* in Alaska. *Courtesy of Insitu Inc. ScanEagle is a product of Insitu.*

improved safety and lower noise levels as compared to using manned aircraft." A week later, responding to an inquiry from the website *sUAS News*, the company confirmed that a ScanEagle the team launched on a second flight ditched in the water after its engine failed.[17] "As you may know, ConocoPhillips and its collaborators successfully completed the first approved commercial use of an unmanned airborne system in US controlled airspace," the company said in an e-mailed statement. "However, during a second test flight, the aircraft experienced engine failure and—as it is programmed to do—aborted the flight into the water. The aircraft was recovered."

AeroVironment and BP claimed credit for the first FAA-approved commercial unmanned aircraft flight over land on June 8, 2014, using the Puma in Alaska. With the first flight, AeroVironment also announced a five-year contract award from BP to provide geographic information system and mapping services at the oil company's Prudhoe Bay oil field on the North Slope.[18] The 13.5-pound, battery-powered Puma, equipped with either a pulsed laser called a light detection and ranging sensor or electro-optical and infrared cameras, transmitted measurements and imagery BP used to generate three-dimensional computerized models of roads, gravel pits, drill pads, and pipelines. It required no runway, catapult, or other infrastructure; hand-launched, the miniature airplane automatically landed within eighty feet of a chosen GPS coordinate. It ultimately belly flopped to the ground, where its wing popped off by design. In August 2014, the Coast Guard and NOAA demonstrated the Puma's potential for oil spill tracking, for the first time launching and recovering one from a ship (the *Healy*) underway in the Arctic Ocean.

Within a year, the *Westward Wind's* achievement had faded as a news angle. Another provision of the FAA Modernization and Reform Act presented a way forward as industry awaited the small UAS rule: Section 333 of the act, entitled "Special Rules for Certain Unmanned Aircraft Systems," gave the secretary of transportation the authority to determine "if certain unmanned aircraft systems may operate safely in the national airspace system" before the FAA completed its slow-going rule-making process. Some familiar names—Amazon, Chevron USA, Delphi Automotive, State Farm, and Yamaha Motor USA—were among the first wave of applicants seeking exemptions to fly UAVs under Section 333, but the Department of Transportation and the FAA granted the first exemptions to six small aerial photo and video production companies endorsed by the Motion Picture Association of America.[19] The FAA's former deputy director of flight standards, John McGraw—then working as a consultant—advised them. The companies each proposed flying miniature, battery-powered helicopters of various configurations: octocopters with eight separate arms and rotors; hexacopters with six; quadcopters with four; or "coaxial quad multirotors" (eight-rotor machines featuring four arms with top and bottom rotors spinning in opposite directions). McGraw recalled when the Motion Picture Association approached him in late 2013, it was unclear which regulatory path the production companies needed to take to use UAVs for aerial filming. The applicants explored several ideas with the FAA, including flying their machines under an experimental research and development certificate. There were "several dead ends." Eventually, they settled on the Section 333 provision Congress had inserted in the FMRA legislation. "We just realized at the end of

The *Westward Wind* was a crab catcher/processor ship converted into a research vessel for service in the Arctic Ocean. ConocoPhillips launched the ScanEagle from its deck to monitor marine mammals in support of its oil and gas exploration activities. *Courtesy of Bill German Outdoors Photography.*

the day it boiled down to an exemption like any other. We spent several months developing what we thought would be successful and filed it with the FAA," McGraw said. "The FAA I think made a reasonable decision to allow it."

The FAA granted the exemptions "for the purposes of scripted, closed-set filming for the motion picture and television industry," and each came with thirty-five conditions attached. Among them, the aircraft was to be operated within the pilot's line of sight and could fly no higher than 400 feet above the ground and no less than 500 feet away from persons not involved in the production. The pilot was to have a private pilot's license with an accompanying third-class medical certificate and be trained and proficient on the machine he or she was operating. All flights required a separate visual observer. The video company was to obtain a COA and also submit a written plan of activities to the local FAA flight standards district office at least three days before scheduled filming.

Transportation Secretary Anthony Foxx, FAA Administrator Michael Huerta, and Dodd—then serving as the motion picture association's CEO and chief lobbyist in Washington, DC—announced the first exemption awards in a teleconference with reporters on September 25, 2014. "Our challenge at the FAA is to integrate unmanned aircraft into the busiest and most complex airspace system in the world, and to do so while we maintain our mission—protecting the safety of the American people in the air and on the ground," Huerta instructed. "We're taking a reasonable and responsible approach. We're introducing unmanned aircraft into America's airspace incrementally and with the interest of safety first." The first companies were required to show their use of unmanned aircraft provided an "equivalent level of safety" to a manned operation and were in the public interest—two bars the FAA and the Department of Transportation determined the companies had passed.

Unmanned, remotely piloted robotic aircraft had flown at the country's margins largely unnoticed for some two decades already, and, like cellular phones, had shrunk in size and evolved with advances in microelectronics, materials, satellite navigation, battery chemistry, and other technologies. Their military reputation preceded them, and the media often miscast them as weapons of war in every circumstance. There was even confusion about what to call them.

CHAPTER 2

DRONE ETYMOLOGY

By the time you read this, the campaign to retire the word "drone" will have been all but lost, undermined by a mainstream media that embraced the descriptor and would not let go. Many other ostensibly more accurate, less objectionable terms were used in its place, trying the patience of the acronym averse. The reader is forewarned that many such terms will be used in these pages. The industry and the FAA preferred UAS—for unmanned aircraft system—because an airplane or a helicopter with no human on board is a component of a larger system; it needs an operator in a ground control station and a radio frequency link to work. The Department of Defense also used UAS, but the Air Force specified remotely piloted aircraft (RPA) to recognize the pilot in the loop. The Navy conducted the historic first "arrested landing" of an unmanned combat air system (UCAS) demonstrator on a modern aircraft carrier in July 2013, catching the Northrop Grumman X-47B with an arresting cable on the USS *George H.W. Bush*. Its next pursuit was to build an Unmanned Carrier-Launched Airborne Surveillance and Strike system. The European Commission was keen to use RPAS, for remotely piloted aircraft system. So too was the International Civil Aviation Organization (ICAO), the United Nations agency based in Montreal that promulgated "standards and recommended practices" for 191 member countries. The 1980s-vintage UAV was still part of the lexicon, but appeared to be fading. "[I]t is no longer aligned with NATO thinking and, in the interests of interoperability, UAV should now be considered a legacy term," Britain's Ministry of Defense advised in a 2013 memorandum.

Whatever the cumbersome acronym, the US government and the unmanned aircraft systems industry considered it better than using the word *drone*, which they deemed to be pejorative. "We call it a system. They're not drones," scoffed Dave Morton, an aviation safety inspector with the FAA. "You fly drones out of Point Mugu (the Navy's sea test range in California). You launch them off a catapult maybe or off a runway. It's programmed to go one way...and maybe the Navy or the Air Force uses it as a target and they shoot it down. That's a drone." In the words of Michael Toscano, then president and CEO of the Association for Unmanned Vehicle Systems International (AUVSI), the industry's largest trade group: "A drone is a dumb bee for most people, or you 'drone on.' It just does what it's programmed to do without thinking. Most people have a negative and incorrect conception of that word." The powerful Aerospace Industries Association stated: "The use of the term 'drone' to describe UAS platforms is a misnomer.

The vast majority of UAS flights are piloted by ground-based personnel…Even when flying pre-programmed routes and missions, real-time pilot intervention is always available." Fully autonomous operation, in which a robotic aircraft would decide for itself how to fly absent any human inputs or control, was still a paradigm shift away.

Britain's ministry of defense rolled out a public relations campaign to correct the record. For the first time in December 2013, the ministry invited journalists to tour its ground control center at Royal Air Force station Waddington, from where RAF pilots flew MQ-9 Reapers over Afghanistan. Simultaneously, then-Defense Secretary Philip Hammond wrote an op-ed article that was published in *The Guardian*. Much of the spirited criticism of RPAS on that side of the Atlantic was "based on genuine misunderstanding or a wild misrepresentation of reality," Hammond complained. "The most basic falsehood, invariably, is the use of the term 'drone,' a name that conjures up images of computer-controlled machines, free from human oversight…In the MOD we call it a remotely piloted air system, because that's exactly what it is. Though physically unmanned, the aircraft is guided and controlled by a team of highly trained people." The headline the newspaper used above the Hammond editorial read: "In defence of drones."[1]

Both the US and British military establishments started experimenting with UAVs around the time of World War I, a little more than a decade after the Wright brothers' first manned flight at Kill Devil Hills, North Carolina. The earliest manifestations the US military developed—the Curtiss-Sperry Aerial Torpedo and the Kettering Bug—were essentially winged bombs that were not destined to return from their missions. They were considered the forebears of modern-day cruise missiles. The appellation drone—in nature a male honeybee that exists solely to mate with the queen and thereafter die—owes its existence to the radio-controlled Fairey Queen aerial target the British Royal Aircraft Establishment developed in the early 1930s. While the Fairey Queen never entered production, the DH.82B Queen Bee that followed was a success. First flown in 1935, the pilotless version of the de Havilland Tiger Moth biplane served the British military as an anti-aircraft gunnery target into the 1940s. The British built some 420 Queen Bees.[2]

According to various accounts, Adm. William Standley, then chief of naval operations with the US Navy, saw Queen Bees demonstrated while attending the second London Naval Disarmament Conference that began in December 1935, with the purpose of reining in naval weapons. Standley was impressed, and upon returning to the United States ordered his service to begin developing unmanned aerial targets.[3] The job fell to LCDR Delmar S. Fahrney, whom the Navy Bureau of Aeronautics assigned to lead its radio-controlled aircraft project. In the semiannual report he submitted in December 1936, "the Officer in Charge of the project initiated the term 'drone' as descriptive of the radio-controlled aerial targets. That designation has persisted and will be used hereinafter," Linwood Howeth wrote in a 1963 book, *History of Communications—Electronics in the United States Navy*.[4]

The Navy officially called the TDN-1 airplane it started developing in 1941 an "assault drone," because the plan was to use it either as a sacrificial guided missile or to drop bombs or depth charges and return to base. The service later commissioned a less expensive type manufactured by Interstate Aircraft—the TDR-1—which saw action in the Pacific Theater during World War II. The Navy's Special Air Task Force flew the twin-engine airplane sans pilot into Japanese bunkers and gun positions at Bougainville in September 1944, and later used it to drop bombs on a nearby island.[5] The use of unmanned aircraft fitted with cameras for surveillance and reconnaissance missions traces to the 1950s. The Army's SD-1 surveillance drone, also known as the Radioplane/Northrop MQM-57 Falconer, was a target drone derivative that operators launched from a platform with booster rockets that was controlled in flight by radio signals and recovered by parachute; it served with the Army and NATO into the 1970s.[6] During the Vietnam War era, the Navy deployed the unmanned QH-50 DASH (Drone Antisubmarine Helicopter) aboard destroyers; another user was Japan's Maritime Self-Defense Force. The Navy's original intention was to fit the QH-50 with torpedoes to ward off Soviet submarines, but the small helicopter served other requirements beyond antisubmarine warfare and was variously equipped with a minigun, grenade launcher, bombs, and other weapons. Some 810 were manufactured before the Navy retired the QH-50 in 1970–1971.

Author and defense industry analyst Steven Zaloga, who started writing about UAVs in the 1980s, found it ironic the modern-day US military blamed the mainstream media for a word that it originally coined. He did not think "drone" conveyed a negative impression. "If you know where it comes from, a lot of the connotations to it disappear," Zaloga offered. "Part of it is that DOD itself is so historically ignorant. They're not even familiar with where the name 'drone' comes from, and they associate it more with the press and popular terms, not realizing that it was for decades the actual term used by the War Department and later by the Department of Defense to refer to remotely piloted aircraft."

September 11 Terrorist Attacks

The terrorist attacks of September 11, 2001, that took down the majestic twin towers of the World Trade Center in New York City, crumbled the west face of the Pentagon, and downed an airliner in a field in southwestern Pennsylvania—killing nearly 3,000 people in all—provided the US government another rationale for using armed drones. Three days after the attacks Congress passed a joint resolution called the Authorization for Use of Military Force (AUMF for short), which President George W. Bush signed into law on September 18. The statute authorized the president "to use all necessary and appropriate force against those nations, organizations or persons he determines planned, authorized, committed or aided" the attacks.[8] One day earlier, the president had signed a memorandum that authorized the CIA "to operate freely and fully in Afghanistan with its own paramilitary teams, case officers and the newly armed Predator drone," Bob Woodward wrote in *Bush at War*.[9]

In the shocked aftermath of a late summer morning gone horribly wrong, the United States adopted a policy of "targeted killings" to hunt down those responsible for the attacks—the terrorist group Al-Qaeda and the Taliban Islamic fundamentalist movement that sheltered it in Afghanistan.[10] Among the principal instruments of what Bush originally called the "war on terror" were the General Atomics MQ-1 Predator fitted with twin laser-guided AGM-114 Hellfire missiles and its larger sibling, the MQ-9 Reaper, capable of delivering missiles and guided bombs.

Unmanned aircraft became known more for what they did than for what they were. It would be a year before the American public became aware of targeted drone attacks, which were commonly thought to have started with a strike that killed six Al-Qaeda operatives in Yemen in November 2002.[11] However, the first "Hellfire Predator" missile strike in combat actually took place a year earlier, on October 7, 2001, the day hostilities

With a maximum gross takeoff weight of 11,700 pounds, the General Atomics MQ-9 Reaper—also known as the Predator B—was capable of delivering four Hellfire missiles, as well as laser-guided bombs and precision munitions. It was larger, heavier, and more lethal than the MQ-1 Predator, which had a 2,250-pound maximum takeoff weight and carried two laser-guided Hellfires. MQ-1 production ended in 2011. *Photo courtesy of the US Air Force.*

began in Afghanistan; journalist Seymour Hersh revealed the basics of the mission in an article *The New Yorker* published in its October 22 issue. Hersh wrote that unnamed military and intelligence officials he interviewed were "crestfallen" that the mission failed to achieve the objective of killing Taliban leader Mullah Muhammed Omar because of "cultural" and bureaucratic obstacles to causing "collateral damage."[12] In the 2014 book *Predator: The Secret Origins of the Drone Revolution*, author Richard Whittle provided a blow-by-blow account of the mission. According to Whittle's

Predator 3034 was an armed prototype the US military and CIA rushed into service over Afghanistan, where it became the first unmanned aircraft to fire a Hellfire missile in combat on October 7, 2001. It was later suspended from the ceiling of the Smithsonian National Air and Space Museum in Washington, DC. *Author photo.*

reporting, the drone behind the first strike—Predator 3034—was an armed prototype the military and CIA rushed into service and launched from a remote airfield in Uzbekistan, near the Afghan border. An Air Force and contractor team stationed at CIA headquarters in Langley, Virginia, controlled the aircraft at altitude by satellite link. For hours, the Predator stalked a convoy of vehicles thought to contain Omar, which stopped at a compound near Kandahar, but US military leaders' consternation over the presence of what could have been a mosque in the Predator's video feed prevented them from acting. Finally, Army Gen. Tommy Franks, watching from US Central Command at MacDill Air Force Base in Tampa, Florida, gave the order to fire—not at the building of interest, but at a nearby truck. All hands watched as the truck exploded from the missile's impact, sending the body of a guard who had been standing next to it spinning into the air. "A new way of waging war had been inaugurated, a new way of killing enemies proven," Whittle wrote. An intelligence analyst attached to the mission at Langley taped to the wall a freeze-frame image of the Predator's video from the moment the missile struck, "a record of the world's first intercontinental drone strike. Soon there would be...many more." Nevertheless, Predator 3034 had a peaceful retirement; it was suspended from the ceiling of the Smithsonian National Air and Space Museum in Washington, DC, as part of the museum's military UAV exhibit that opened in 2008.[13]

Drone Strikes Escalate

Basing its numbers on wire service, regional newspaper, and network television reports, the Washington, DC, based New America Foundation—a nonprofit public policy institute—found that President Bush presided over one drone strike in Yemen and forty-eight in the tribal regions of Pakistan from 2002–2009. In his first four years in office through 2012, President Barack Obama presided over 352 strikes—seven times Bush's total in those countries. The Bureau of Investigative Journalism, a philanthropically funded journalism shop based at City University London, tallied fifty-two strikes under Bush and 300 under Obama during the same periods in Pakistan alone. Within months, Obama's ramp-up intensified a debate over unintended civilian casualties of drone strikes. "An important factor in the controversy over drones is the widespread perception that they kill large numbers of Pakistani civilians," the foundation said in an October 2009 analysis. "Some commentators have asserted that the overwhelming majority of casualties are citizens." In a later policy paper christening 2009 as "The Year of the Drone," the foundation reasoned that 32% of the estimated 830 to 1,210 people killed by drone strikes in northwest Pakistan since 2004 were civilians.[14] In October 2013, the non-governmental organizations Human Rights Watch and Amnesty International jointly released investigative reports they had conducted of targeted killings in Yemen and Pakistan, respectively. Both organizations alleged the United States had unlawfully killed civilians.[15] That month, two human rights experts the United Nations appointed as "special rapporteurs" to study the role of drone strikes in counter-terrorism operations and extrajudicial killings doubted their legality under international law and called upon user states—namely the United States, but also Britain and Israel—to be "transparent" in their use of drones and to investigate allegations of civilian casualties.[16] Human Rights Watch later published an investigation into a December 2013 drone strike on a convoy of vehicles in Yemen that it determined was a wedding procession that may have included al-Qaeda members, but most of those killed or wounded were apparently civilians.[17]

Obama's administration had built an "extensive apparatus" to conduct drone strikes involving dozens of facilities and clandestine bases in at least six countries, *The Washington Post* reported in December 2011. "[N]o president has ever relied so extensively on the secret killing of individuals to advance the nation's security goals," the *Post* said. Senior administration officials the newspaper interviewed on the condition of anonymity "said the escalating number of strikes has created a perception that the drone is driving counterterrorism policy, when the reverse is true...Nevertheless, for a president who campaigned against the alleged counterterrorism excesses of his predecessor, Obama has emphatically embraced the post-Sept. 11 era's signature counterterrorism tool." But apart from human rights advocates and others in the United States raising concern, the level of public debate remained "muted," the *Post* added.[18] During the 2008 presidential campaign that brought the charismatic Illinois senator to power, "many anti-war activists threw themselves into the election of Barack Obama, only to discover that the peace candidate had morphed into the war president," Medea Benjamin, co-founder of the antiwar group Code Pink, wrote in the 2012 book *Drone Warfare, Killing by Remote Control*. The US government was prosecuting the wars in Afghanistan and as of March 2003 in Iraq with smart bombs, laser-guided missiles, and "a newfangled kind of unmanned planes called drones." It had spent billions of dollars on "building a better flying death robot," she said. "Members of the US peace community watched in horror as these snipers in the sky started spreading from Afghanistan and Iraq to Pakistan, Yemen, Somalia, the Philippines and Libya."[19]

The Obama administration justified its practice of targeted killings in a series of major policy speeches that began in March 2010.[20] It was not until April 30, 2012, that it officially acknowledged using drones to carry out those killings, in a speech then-counterterrorism advisor John O. Brennan delivered at the Woodrow Wilson International Center for Scholars in Washington, DC. Seven minutes into the speech, the elfin, middle-aged activist Medea Benjamin stood up and, facing away from Brennan toward the back of the room, announced she was speaking out on behalf of the "innocent victims" of drone strikes in Pakistan, Yemen, and Somalia. "They deserve an apology from you, Mr. Brennan," she shouted. "How many people are you willing to sacrifice?" As seen in a YouTube video, Benjamin ignored a woman tugging at her jacket and continued her monologue. Soon a burly policeman appeared, snatched her from under the armpits, and hauled her out of the room. She was last heard yelling "shame on you!" as the door closed—an incident C-SPAN filmed but the Wilson Center edited from its official record of the speech. With Benjamin evicted, Brennan carried on with the address. "Yes, in full accordance with the law, and in order to prevent terrorist attacks on the United States and to save American lives, the United States government conducts targeted strikes against specific al-Qaeda terrorists, sometimes using remotely piloted aircraft, often referred to publicly as drones," he declared. "And I'm here today because President Obama has instructed us to be more open with the American people about these efforts." The strikes were legal—allowed by the 2001 AUMF and international law—Brennan argued, and were also ethical and wise. "There is, of course, no such thing as a perfect weapon, and remotely piloted aircraft are no exception," he said. "As the president and others have acknowledged, there have been instances when, despite the extraordinary precautions we take, civilians have been accidentally injured, or worse, killed in these strikes. It is exceedingly rare, but it has happened. When it does, it pains us, and we regret it deeply, as we do any time innocents are killed in war."[21]

Less than a year later, Brennan would figure in another monologue, this one of epic proportions. In January 2013, Obama nominated the hard-headed, second-generation Irishman and veteran spy to serve as CIA director in place of David Petraeus, the former army general and Iraq war commander who had resigned months earlier over an extramarital affair. Code Pink protesters repeatedly interrupted Brennan's nomination hearing before the Senate Intelligence Committee in February, causing Senator Dianne Feinstein, the committee's chairwoman, to clear the room before resuming. A prominent line of questioning by senators concerned the government's exercise of due process under US law when targeting American citizens overseas. The issue arose after the CIA's takedown of Anwar al-Awlaki, a Yemeni-American cleric, and Samir Khan, another American citizen and al-Qaeda operative, in Yemen in September 2011. The government believed the fiery preacher al-Awlaki inspired and abetted the Nigerian "underwear bomber" involved in the failed plot to explode an airliner over Detroit on Christmas Day 2009, and that he also played a role in al-Qaeda's October 2010 attempt to destroy cargo airplanes using plastic explosives hidden in printer cartridges. A drone strike killed al-Awlaki's teenage son in Yemen within weeks of his father's death, although reportedly he was not the intended target. As the Senate mulled Brennan's nomination, Rand Paul, the libertarian Republican senator from Kentucky, staged a thirteen-hour filibuster on March 6 to delay it from voting. He demanded the Obama administration answer whether it had authority "to use a weaponized drone to kill an American not engaged in combat on American soil." After the filibuster ended, the Senate voted 63–34 to confirm Brennan as CIA director on March 7. In a letter dated the same day, Attorney General Eric Holder provided Paul with a one-word answer to his question: "no."[22]

On May 23, 2013, Obama weighed in with a speech on counterterrorism strategy at National Defense University, located on the grounds of Fort McNair in Washington, DC. He argued US actions in prosecuting the war against al-Qaeda and its offshoots were legal. "We are at war," the president said, "with an organization that right now would kill as many Americans as they could if we did not stop them first. So this is a just war—a war waged proportionally, in last resort and in self-defense." He addressed the government's use of Predators specifically:

> [M]uch of the criticism about drone strikes, both here at home and abroad, understandably centers on reports of civilian casualties. There's a wide gap between US assessments of such casualties and nongovernmental reports. Nevertheless, it is a hard fact that US strikes have resulted in civilian casualties, a risk that exists in every war. And for the families of those civilians, no words or legal construct can justify their loss. For me, and those in my chain of command, those deaths will haunt us as long as we live, just as we are haunted by the civilian casualties that have occurred throughout conventional fighting in Afghanistan and Iraq.

Congress was also involved—it had authorized the use of force by passing the AUMF and was briefed on "every strike that America makes," including the killing of Anwar al-Awlaki.

The president also spoke about his desire to close "Gitmo," the military prison camp the Bush administration established at Guantanamo Bay Naval Base, Cuba, to house suspects captured in the war on terror. As he did, a woman in the audience repeatedly interrupted him, shouting that as commander-in-chief "you can close Guantanamo Bay." The normally contained Obama for a moment appeared agitated, his eyes ablaze. "Let me finish, ma'am. Part of free speech is you being able to speak, but also, you listening and me being able to speak," he scolded the heckler, as the audience applauded. The woman continued shouting about Guantanamo prisoners, CIA drone strikes, and "thousands of Muslims that got killed" until security escorted her from the room. "I'm going off script, as you might expect here," Obama said once peace was restored. "The voice of that woman is worth paying attention to. Obviously, I do not agree with much of what she said, and obviously she wasn't listening to me in much of what I said."[23] The woman was Medea Benjamin, who afterward told *The Huffington Post* that a person she would not identify had given her an invitation to the speech. While several reporters recognized her as they filed into the event, Benjamin tried to remain inconspicuous by talking on her cell phone. "I must say, I do really appreciate that I live in a country where if you interrupt the president you don't get beaten and tortured and thrown inside a prison for a year," she told the online news site.[24] Benjamin was not unaware of the potential consequences of her activism; several months earlier she had participated in a risky "peace delegation" to Pakistan to protest drone strikes alongside the prominent politician and former cricketer Imran Khan. In 2014, she charged that Egyptian police broke her arm after detaining her in Cairo.

With offices on both coasts, Code Pink—a name that played on the Bush administration's color-coded security alerts—became the *bête noire* of the US government and unmanned aircraft systems industry. In addition to disrupting speeches and congressional hearings, it staged "creative protests" featuring activists dressed in cardboard and paper maché drone costumes and "die-ins" at which they draped themselves in sheets splattered with fake blood. It proclaimed "citizens arrests" in public venues, crashed book signings, and laid siege to the homes of government and corporate executives in tony enclaves like Georgetown and La Jolla. "Naming and shaming" the companies involved in and profiting from the manufacture of weaponized drones was a preferred tactic.

The antiwar group Code Pink staged "creative protests" to show its displeasure with drone missile strikes. In May 2014, in front of the White House, the group held a mock wedding ceremony that was interrupted by a figurative drone strike, depicting what apparently happened to a wedding in Yemen. *Author photo*.

Elsewhere other groups waged drone warfare. "You may not have heard of the 'Creech 14,' but they have a special place in the heart of the anti-drone movement," Benjamin wrote in her book. On April 9, 2009, the group entered Creech Air Force Base in Indian Springs, Nevada, about thirty-five miles northwest of Las Vegas, where Air Force pilots flew Predators and Reapers in support of "overseas contingency operations" in Afghanistan and Iraq. The activists were arrested for trespassing and spent a couple days in jail; some two years later a judge found them guilty, sentenced them to time served, and told them to "go in peace." In January 2010, a group formed by anti-war activist Cindy Sheehan, who lost her son in the war in Iraq, demonstrated outside of CIA headquarters in suburban Virginia and unfurled a banner reading "Drones Kill Kids" at the Smithsonian's military UAV exhibit. That May, Code Pink staged a "somber vigil" at the La Jolla residence of General Atomics CEO Neal Blue and the following day protested at the company's San Diego headquarters. In April 2011, thirty-eight activists were arrested after conducting a die-in at Hancock Field Air National Guard Base near Syracuse, New York, from where the 174th Attack Wing operated MQ-9s over Afghanistan. This was the "Hancock 38." As he had earlier during the trial of the Creech 14, former Johnson administration attorney general Ramsey Clark testified in support of the Hancock protesters.[25] By 2012, the group San Diego Veterans for Peace was holding weekly protests outside General Atomics' corporate headquarters. The "No Drones Network" of protest websites recorded demonstrations in eighteen states. The Nevada Desert Experience, Voices for Creative Nonviolence, Nevada County Peace Center, Code Pink, and other groups held monthly demonstrations outside US air bases. "It's hard for folks living or commuting past the bases to miss the protesters' messages," Benjamin wrote. "Between 2010 and 2012 the movement grew tremendously…The American public had begun to wake up from its drone-induced slumber."[26]

Code Pink's Washington, DC, headquarters was a tan, wood-framed house on a tree-lined street at the city's urban edge. There was a white picket fence with multi-color points and a wraparound porch with a drawing board and some faux gravestones stacked on one side. With a knock Medea Benjamin came to the front door, which was painted pink. The walls inside the house were painted in various pastel colors and decorated with posters and signs made of construction paper, with messages such as "Arms are for Hugging"—the motto of a campaign to ban assault weapons—and "End US military aid to Israel," another of the group's causes. On this rainy weekday morning, three twenty-something women and a man were arrayed around computers at the dining room table. They looked up to say hello, but otherwise stayed busy behind their laptops. Benjamin was petite, with chestnut-colored, shoulder-length hair and eyes shaped like almonds. No doubt intrigued that an aviation trade press reporter had come calling, she readied a cup of coffee in the kitchen. She discovered the milk in the group's refrigerator was spoiled so instead offered flavored creamer.

In a sense, the world had come full circle since Benjamin, Jodie Evans, and other activists founded the anti-war group in 2002. US warplanes and Predators had returned to the skies over Iraq to stem the advance of a barbaric movement calling itself ISIS, or the Islamic State. In *Drone Warfare*—essentially a 225-page indictment of drone strikes—Benjamin allowed that "not all uses of unmanned aircraft are bad," as some drones could serve useful purposes, such as assisting firefighters in battling forest fires, or with their cameras bear witness for environmental and human rights groups. Even Code Pink had one: an AR.Drone quadcopter that it arguably flew in contravention of FAA rules when

Code Pink co-founder Medea Benjamin disrupted speeches by major figures in the Obama administration, including President Obama himself. She was a prominent leader of the anti-drone movement. Benjamin ackowledged drones could be useful commercially; what Code Pink opposed was their use for "repressive" purposes. *Author photo.*

the group demonstrated at Blue's residence. But it wasn't such purposes that fueled the drone boom; rather, it was "state-sponsored assassinations and semi-covert wars," she wrote at the time. "I would like to modify that a little now," Benjamin said in retrospect, as we sat on the living room couch. "I think there are, certainly in terms of the smaller drones, lots of commercial uses for them and just in terms of sheer numbers we'll see a lot more drones being used for private and commercial purposes. Our position is we don't want drones used for killing, for spying on us and for repressive purposes."

Robotic aircraft had already graduated from activists' niche cause to cultural phenomenon. The CIA's use of the Predator for targeted killings was depicted in the 2005 film *Syriana*, starring George Clooney and Matt Damon—in this case the agency eliminates an Arab prince who had run afoul of the US government and oil industry. In the 2011 television series *Homeland*, Marine Sgt. Nicholas Brody, held prisoner by al-Qaeda for eight years in Iraq, turns terrorist against his own country. The transformation happened after a US drone strike killed his captor's young son, whom Brody had taught English and come to love. An armed Predator pursues agent Aaron Cross in Alaska in the film *The Bourne Legacy*, released in 2012. In January 2013, public television broadcaster PBS aired "Rise of the Drones," an episode of the *NOVA* science series. October saw the debut of filmmaker Robert Greenwald's documentary *Unmanned: America's Drone Wars*, which examined the impact of civilian deaths caused by drone strikes. Greenwald helped bring some of the victims to Washington to testify before Congress, the first time that had been done.[27] The movie *Drones* was a morality play performed entirely in a trailer at what was supposed to be Creech Air Force Base. An MQ-9 Reaper crew—sensor operator Lt. Sue Lawson, a general's daughter, and jaded airman Jack Bowles—clash over taking out a suspected al-Qaeda terrorist in Afghanistan who is surrounded by his family. Lawson initially opposes the mission because it will kill children, but ultimately the two switch convictions. *Good Kill*, a 2014 film, presented the protagonist played by Ethan Hawke with a similar moral conflict over killing terrorists from an air-conditioned trailer thousands of miles removed from the scene. In a more positive light, Insitu's ScanEagle made a cameo appearance in the 2013 film *Captain Phillips*. Starring Tom Hanks, the movie was the true story of the Navy's dramatic rescue of the captain of the cargo ship *Maersk Alabama* after Somali pirates took him captive and fled aboard one of the ship's lifeboats. The actual ScanEagle used in the rescue mission was placed on permanent display at The Museum of Flight on the grounds of Boeing Field near Seattle. *The November Man*, a 2014 film starring Pierce Brosnan, featured a small surveillance multi-copter that a videographer might buy, bringing notoriety to a new order of drones.

Among written works Richard Clarke, who served as national counterterrorism coordinator under presidents Clinton and Bush, published a thriller *Sting of the Drone* in 2014. The novel opens with a CIA covert operation that employs two "stealth mini-drones" to assassinate a cartel leader with terrorist ties at an elegant hotel in Vienna—and this from a man who well knew what the US government was capable of. Whittle's Predator introduced readers to Abe Karem, the brilliant but difficult Israeli engineer who designed the iconic drone for which the book is named, and the swashbuckling brothers Neal and Linden Blue, whose company General Atomics built it. Journalist and author Dan Fesperman's novel *Unmanned* raised "many concerns about drones, or UAVs," *The Washington Post* said in a review. "It is suggested that their use in Afghanistan has been a testing ground for military contractors who hope to dominate the future domestic use of drones, which promises huge profits and raises serious questions about the government's ability to spy on its citizens."[28]

A search of the ProQuest NewsStand database using the terms "drone," "unmanned," and "aircraft or aerial" to strain out entomological references to drones, revealed fifty-two news articles among major and other English-language newspapers and magazines in the year 2000. This followed a decade in which the US military and allied countries used unmanned aircraft extensively in the Gulf War of 1990–1991 and during the Balkan wars in the mid-1990s, where the US first deployed the Predator. The same word search for the eventful year 2001 produced 501 hits—a tenfold increase. The number of articles referencing drones increased to 648 in 2002 and 747 in 2003, as the US went to war in Afghanistan and Iraq. Over the next four years the hits declined, averaging 450 a year. They spiked to 781 in 2008, and thereafter spiraled each year to 4,144 returns in 2013. The February 11, 2013, cover of *Time* magazine exemplified the mainstream media's portrayal of drones as they became more of a prospect on the home front. Against a stormy-grey background it featured a "photo illustration" of an armed Predator hanging menacingly over a residential dwelling. "RISE OF THE DRONES," the headline shouted in capital letters. "They are America's global fighting machines. What happens when they're unleashed at home?" That year, the small town of Deer Trail, Colorado—an hour east of Denver by car—made national news when one of its residents proposed an ordinance that would authorize the town to issue hunting licenses and award bounties to anyone who shot down an intrusive drone—in particular one with federal markings. After various delays, residents voted down the proposal the following April Fool's Day and also dispatched the mayor, one of its proponents.[29] In May 2013, the Aerospace Industries Association issued a report titled "Unmanned Aircraft Systems: Perceptions and Potential," stating that its purpose was to define and "demystify" unmanned aircraft. "As with any emerging technology, public opinion regarding these systems often begins in the imagination, and may harden into myth through misconception, popular culture and an inability to imagine the non-military benefits of a platform that has traditionally been used for national defense," the report argued. "As a growing activist community responds to frequent headlines about the military use of unmanned aerial vehicles…

there has been little acknowledgment of UAV use for humanitarian, disaster response, search and rescue and other life-saving applications."

Opinion surveys from the time revealed that Americans were not concerned about Hellfire missile strikes; rather, they were bothered by what drones could do with the ubiquitous camera turrets hanging from their noses and bellies—spy. The AIA sponsored an online survey *The Christian Science Monitor* conducted of its readership in June 2013 to measure support of non-military uses of unmanned aircraft. Of 4,895 respondents from the US and elsewhere, 54% supported or strongly supported increased non-military use of UAS "if their issues of concern are addressed." The top two issues were individual privacy and the safety of people and property on the ground. Non-US respondents were more supportive than their American counterparts, who were more concerned about privacy issues.[30] The University of Oklahoma Center for Risk and Crisis Management conducted an Internet survey in January 2014 that produced responses from 1,364 participants. According to a summary the university provided the author, key findings suggested that "potential uses of UAS will be conditioned by growing privacy concerns and suspicions of government intrusiveness...Perceptions are that government is overemphasizing security at the cost of privacy and other liberties. To the extent that UAS becomes associated with further erosion of privacy, public opposition is expected." Around the same time, the Pew Research Center and *Smithsonian* magazine surveyed 1,001 adults by mobile phone and landline on their attitudes toward various technological advances, including those in bioengineering, wearable computing, and unmanned aircraft. The results: 63% of respondents thought "it would be a change for the worse if 'personal and commercial drones are given permission to fly through most US airspace.'"[31]

Clearly the unmanned aircraft systems industry in the US had a perception problem, one that it struggled to address. In August 2011, Medea Benjamin "hijacked" AUVSI's pre-convention press briefing at the National Press Club in downtown Washington, DC. "As soon as we recognized Benjamin sitting in front of us, we expected sparks to fly," *FlightGlobal's* Stephen Trimble wrote in a blog post. "Honestly, we were a little disappointed by how quickly the protest was subdued. After this one-minute outburst, Benjamin sat quietly while we asked AUVSI's panelists our normal questions about cost reductions, export prospects and operational doctrine."[32] The following year, Benjamin claimed her effort to book a room at the Las Vegas Convention Center to discuss *Drone Warfare* during AUVSI's Unmanned Systems Conference was thwarted, as was her attempt to register for the conference as a journalist. When she registered on-line as an attendee and arrived to retrieve her badge, police surrounded her and threatened her with arrest. Other known activists met similar receptions.[33] Code Pink's Alli McCracken disrupted a keynote speech US Army Lt. Gen. James Barclay gave at AUVSI's 2013 conference in Washington, DC. Before being ejected, she unfurled a banner and told conference attendees they had "the blood of innocent children on your hands."[34] The antiwar group staged its creative protests in front of the Washington and Las Vegas convention centers, obliging the suits attending those events to walk by a gauntlet of counterculture demonstrators. "I think in the beginning they were kind of surprised that there would be opposition," Benjamin later mused.

Skirmishing with battle-hardened peace activists was one thing; winning over an American public with an ingrained mistrust of government dating to the Revolutionary War was something else. Industry proponents tended "to get in a protective crouch" when reporters approached, University of Oklahoma professor Hank Jenkins-Smith told a meeting of AUVSI's Oklahoma chapter in March 2013. "The problem is that adopting that strategy works against them every time. If you approach journalists as if they're the enemy, they'll pick up on it." Chapter president James Grimsley said the privacy issue arising from government and police use of surveillance drones presented the trade organization with a serious challenge. "We're dealing with fear, and fear is a very real emotion," he said. "We don't want to combat it; we have to educate. We desperately need to think about how people perceive what we do. We have to be patient and we have to take the high road to educate."

At AUVSI's 2014 conference in Orlando, the French consumer electronics manufacturer Parrot—a first-time exhibitor—reserved a prominent space on the show floor to demonstrate its new AR.Drone 3.0 quadcopter named "Bebop"—essentially a flying fourteen-megapixel "fish-eye" camera—and other rolling, jumping, and flying curiosities from its minidrone product line. The FAA's Jim Williams was among spectators as a French millennial piloted the buzzing drone to and fro using a tablet computer. Parrot's presence was emblematic of the annual trade show's gradual evolution to a civilian commercial event. In years past, the Unmanned Systems Conference was peppered with brass and camouflage; now it featured sophisticated toys alongside military exhibits and uniforms were harder to spot among the 8,000-odd attendees. Still, AUVSI's leadership reflected the sponsor's military heritage dating to 1972, as the United States began disengaging from the Vietnam War. John Lademan, Northrop Grumman director of strategic planning, was chairman of the executive committee of the 501(c)(6) non-profit organization, serving with representatives of two other aerospace defense companies: Airbus Group and L-3 Airborne Technologies. Gene Fraser, Northrop Grumman's engineering sector vice president, was

French consumer electronics manufacturer Parrot featured its new Bebop flying camera drone at the AUVSI annual conference in Orlando in 2014. Parrot's presence was emblematic of a shift in the trade show from mainly military to a civil and commercial-oriented event. *Author photo.*

treasurer. Peter Bale, principle with Sentinel Robotic Solutions—a small defense-oriented business—was the immediate past chairman. "This technology has emerged from the military side of the house, so most of our membership in the past has been military and your board should be representative of your membership," explained CEO Michael Toscano. AUVSI counted among its directors representatives of the farm and construction machinery manufacturer John Deere and German automotive supplier Continental, which was pursuing self-driving cars. After all, the association was more than about unmanned aircraft—it also represented the interests of undersea and ground robotic vehicle developers. "Now we're recognizing, just like we have with the computer, with GPS, that it's sliding into the commercial side," Toscano continued. "As the technology becomes available, we will embrace that and modify the board accordingly."

AUVSI resided on the fourth floor of an office building in the northern Virginia suburb of Arlington, less than five miles from the Pentagon, near where Route 395 intersects Quaker Lane. On a gloomy winter day some months before the Orlando conference, the National Weather Service issued a tornado watch for the Washington, DC, metropolitan area. Toscano apologized after arriving minutes late for an appointment. He was dressed in shirtsleeves and a bright yellow tie that offset his dark black eyebrows and the hair that covered two-thirds of his head. Something unexpected had happened earlier that week; hackers had breached AUVSI's email system and he was busied calming concerned members and instructing them to change their passwords.

Toscano came to AUVSI straight from the Pentagon. He was working in the building the day in September 2001 when American Airlines Flight 77 slammed into its western face, killing all aboard the airplane and 125 in the Pentagon, including people who were

Michael Toscano led the Association of Unmanned Vehicle Systems International for six years during a transformational time for the UAS industry in the United States. Before that, he worked for three decades for the Department of Defense and came to AUVSI directly from the Pentagon. *Author photo.*

his friends. Years later, he wore a lapel pin that commemorated that event. Before AUVSI's directors appointed him to lead the association in September 2008, he was program manager of nuclear security research and development in the Office of the Deputy Assistant Secretary of Defense for Nuclear Matters—he was involved in "nuke security," Toscano helpfully summarized. During his three-decade career with the Department of Defense, he also coordinated the military's Joint Robotics Program to research, develop, acquire, and field unmanned ground vehicles. An engineer by training, he was passionate about robotics. "By the way, if you haven't realized, I don't 'drink the Kool-Aid.' I bathe in the Kool-Aid," he interjected at one point. "I am all in on this technology. I recognize that it can save lives." Within months of the interview, he informed AUVSI's staff that he planned to retire at the end of the year.

Unmanned aircraft had been given a bad rap in the media and were not responsible for every missile that rained from the sky, Toscano asserted. "If it bleeds it leads," he said, sharing his understanding of the way the media operated. "Every time you hear the words 'drone strike,' what does that mean? Whether it's launched from an F-15, an F-16, an F-18, whether it's launched from an Apache helicopter, whether it's launched from whatever—there's a platform. The platform is just a truck, and it carries a mission payload or in this particular case a weapon. So they're not drone strikes, they're air strikes." In the fog of war, robotic aircraft served as a convenient scapegoat, he said, continuing the point. "I've been told that everything that explodes in Afghanistan or Iraq is called a drone strike. They have IEDs that explode. They have howitzers that are launching munitions, and if it explodes they call it a drone strike. They don't know where it came from but it had to be a drone because we didn't see it. There's a lot of misinformation that goes on." Similarly, unmanned aircraft were misrepresented on the home front. Toscano recalled a study in which testers presented people with photos of various unmanned aircraft and asked them to choose the photo that most signified a drone. "More than 80 percent of the people picked a Predator with a Hellfire missile," he reported. "The news will do a piece on using 'drones' for agriculture or something else and they show a Predator. It's the fear factor, it's the sensationalism, it's the Hollywood, it's the Predator, it's the RoboCop. That's the mentality."

The mention of Code Pink elicits a grin from Toscano. "I know Medea," he said—though both told the author they had never sat down to talk. "I have no problem with anyone expressing their views and rights as a US citizen. That is their prerogative, that is their right do so. I just think that they're misinformed and are missing the issue. If they're opposed to war, that's their prerogative to do so. From that standpoint, I can understand. The difficulty I have is when you don't get the facts right, when you don't get the truth, when you don't understand what you're truly talking about."

Toscano refused to concede that the battle over the word drone was lost in the mainstream media. He pointed to the terminology the FAA and ICAO used for unmanned aircraft systems. "Those are the two authoritative entities. As long as both of those keep using the words UAS and RPAS then that's what we're going to call them," he declared. But it was becoming a lonely battle. In March 2013, the Senate Judiciary Committee invited several witnesses, including Toscano, to testify at a hearing titled "The Future of Drones in America: Law Enforcement and Privacy Considerations." When it came his turn to speak, Toscano remarked that the hearing provided an excellent opportunity to address misconceptions about the technology, such as the use of the word drone, a "hostile connotation" for something the industry called an unmanned aircraft system.

Unlike military machines, the "UAS" local police would use would be small—weighing less than five pounds—and subject to strict FAA regulations, he assured. But before Toscano could finish his statement, the committee's chairman, Vermont Democrat Patrick Leahy, cut him off. "I appreciate you telling us what we should call them, but I think you leave that decision to us. We will decide what we will call them," Leahy admonished. "You call them whatever you would like to call them."

Later that year, at the Unmanned Systems Conference in Washington, DC, media outlets derided AUVSI's choice of a Wi-Fi password for the press room—"DontSayDrones"—as a laughable attempt at persuasion. The following year in Orlando, the association sought to convey a positive message; the Wi-Fi password for the press room was: "SavingTimeMoneyLives."

CHAPTER 3

PUBLIC UAVs

By the time targeted killings overseas made drones a household name, the FAA had been authorizing public universities, federal agencies, and police forces to fly them for about two decades. The early approvals for domestic applications were ad-hoc: FAA local or regional officials authorized flights by issuing waivers for what were considered "noncomplying" operations, or those that did not conform with existing regulations, such as an airshow or parachute jump. FAA Order 7610.4 ("Special Operations") covered procedures for military flights by the Department of Defense in domestic airspace. Faced with an increasing number of requests from civilian entities, the FAA formed an aviation regulatory advisory committee in 1991 to recommend rules specific to UAVs. In September 1992, the agency issued a memorandum to its regional air traffic and flight standards division managers laying out an "Interim UAV Policy" they were to follow as it promulgated rules. "The proponents are interested in a variety of UAV applications. For example, interests vary from the need to test UAVs that are under development to actual applications such as drug interdiction, upper air sampling, aerial photography, border patrol, etc.," stated the memorandum, which was signed by William Pollard, the FAA's associate administrator for air traffic, and Anthony Broderick, associate administrator for regulation and certification. "To ensure standard application" of the interim policy, the memorandum instructed that regional managers were allowed to approve non-DOD unmanned aircraft operations only in restricted or warning areas. They were to forward all other requests to the Air Traffic Rules and Procedures Service office at FAA headquarters in Washington, DC, for processing.

Climate research drove early civil government applications. The Department of Energy introduced UAVs as part of its Atmospheric Radiation Measurement program in the early 1990s. The ARM-UAV program mated scientific instruments with unmanned aircraft to study solar radiation and its interaction with clouds. The program started in November 1993, with engineering test flights at Edwards Air Force Base, California, of the Gnat-750, a precursor of the Predator developed by Abraham Karem's Leading Systems Inc. before General Atomics acquired the company. The first test series was followed by a second set of flights from Blackwell Tonkawa Airport, Oklahoma, in April 1994. These represented the program's first science flights in unrestricted airspace and required that a manned DHC-6 Twin Otter turboprop accompany the Gnat-750 as a chase plane. The ARM-UAV effort also used the NASA-developed Altus II derivative of the Predator A for atmospheric research.[1]

In the absence of standards for the design, manufacture, certification, and operation of unmanned aircraft, the DOE produced its own "UAV Airworthiness Interim Guidance" manual in December 1994—later amended—that drew upon existing FAA regulations and advisory circulars. A source presented a copy of the manual, which was contained in a binder, to the author. "We believe that UAV operations should be as safe as manned operations insofar as they should not present or create a hazard to persons or property in the air or on the ground greater than that created by manned aircraft of equivalent class or category," said a memorandum accompanying the document. "These criteria are particularly important since the proposed Atmospheric Radiation Measurement UAV program will be conducted not in restricted areas, but in the National Airspace System, over populated areas."[2]

NASA's Environmental Research Aircraft and Sensor Technology program (1994–2003) aimed to develop a family of remotely or autonomously operated UAVs and sensors for high-altitude, long-endurance Earth science and environmental missions. Under ERAST, the space agency supported General Atomics in developing the Altair, a civil variant of the enlarged Predator B with an extended wingspan for NASA's purposes. The program also flew the Altus II, the Aurora Flight Sciences' Perseus, the AeroVironment Helios Prototype, and the manned, tandem-wing Proteus aircraft built by Scaled Composites. The solar-powered Helios Prototype—a flexible flying wing—set a world altitude record in August 2001, by ascending to 96,863 feet—more than two miles higher than any non-rocket-powered aircraft had flown.[3] In May 2004, NASA, the FAA, the Departments of Defense and Homeland Security, and an industry grouping called UNITE (the UAV National Industry Team) launched the Access 5 Project to develop standards and procedures that would underpin routine access to the airspace system, initially by high-altitude, long-endurance unmanned aircraft.[4] The space agency committed $103 million to the effort over five years and the UNITE team of AeroVironment, Aurora Flight Sciences, Boeing, General Atomics, Lockheed Martin, and Northrop Grumman pledged $36 million in funding or in-kind services, but within two years, NASA cancelled the project, focusing its resources on space programs.

The National Oceanic and Atmospheric Administration joined with NASA in the spring of 2005 to demonstrate the Altair, fitted with a suite of scientific instruments for "oceanic and atmospheric research, climate research, marine sanctuary mapping and enforcement, nautical charting and fisheries assessment and enforcement" of the Channel Islands National Marine Sanctuary off California. Separately, that September, General Atomics announced that it obtained for the Altair the nation's first experimental airworthiness certificate the FAA granted to a company for an unmanned aircraft: a one-year authorization allowing it to fly the UAV at higher altitudes in an expanded geographic area.[5]

A turning point for private industry's ability to fly unmanned aircraft came on September 16, 2005, when the FAA issued a policy memorandum under the heading, "Unmanned Aircraft Systems Operations in the US National Airspace System-Interim Operational Approval Guidance," providing guidance its branches should use in evaluating whether to grant certificates of authorization or waiver to public agencies seeking to fly unmanned aircraft. Among the agency's requirements, COA applicants were obliged to show their aircraft came with an FAA airworthiness certification, or a statement specifying that Department of Defense airworthiness certification criteria contained in the MIL-HDBK-516 handbook were used to certify the machine. The memorandum, signed by

Flight Technologies and Procedures Division manager John McGraw, plainly stated that "COA applications for civil (unmanned aircraft) operations will not be accepted."[6]

McGraw joined the FAA in 1995. He became manager of the flight technologies division—a branch of the Flight Standards Service—in 2002. Explaining the rationale for the memorandum, he said:

> From the time I came in, (from) 2002 until 2005, we found a wide variance in how these aircraft were being authorized to fly, and how particularly the field offices were handling applications—everything from field offices that thought it was OK to just sign off and let them fly to field offices that were saying 'absolutely not, there's no chance, you can't do it.' We wanted to put out some guidance that told our workforce: this is how you handle these. We didn't think it was a shift in policy; we thought it was what the policy should have been all along.

Until then, the agency had approved some COAs for private companies that did not have public sponsors. That would no longer be the case.

NOAA listed sixteen projects that it conducted from 2005 to 2012 using unmanned aircraft.[7] In addition to the Altair, the models it tested included the Aerosonde, which in 2005 became the first UAV flown into the eye of a hurricane, adding to a list of achievements that included the first unmanned trans-Atlantic flight. NOAA also flew the Office of Naval Research-funded Silver Fox and Manta, the Insitu ScanEagle, the Northrop Grumman Global Hawk, the AeroVironment Puma, and the Aeryon Scout. "We have moved on (from) a specific science topic of climate change to something very much broader, which is 'environmental intelligence.' [W]hat we're looking at specifically for UAS is how can we better monitor different aspects of the climate continuum, and that's high-impact weather monitoring, polar monitoring and marine monitoring," Robbie Hood, director of NOAA's UAS program, told the Unmanned Systems Conference in 2014.

By 2015, the environmental agency planned to take over the hurricane-forecasting mission NASA flew using a pair of Global Hawks stationed at Wallops Island, on the Virginia coast. NOAA's three-year, $15 million project envisioned the aircraft releasing parachute-equipped "dropsondes," tubular devices the National Center for Atmospheric Research developed to measure changes in hurricane intensity. NOAA also planned to make more use of the hand-launched Puma for its mission monitoring marine sanctuaries. "I think the biggest accomplishment in NOAA is that we're starting to use UAS more regularly," Hood related. "We're still not making a lot of purchases yet, because as an agency we have to figure out what is the optimum observing system between satellites, weather radars, weather balloons, and ocean buoys. How do we bring UAS in the mix and make sure that they are providing the best science value in an operationally feasible and cost-effective way? Our scientists now are aware of UAS. It's no longer a question of does NOAA need UAS—that question's been put to bed."

Science agencies weren't the only agencies using unmanned aircraft. The mission of the Bureau of Land Management—an agency of the Department of the Interior—was to manage and conserve 245 million acres of public lands, as well as 700 million acres of "sub-surface mineral estate" underlying federal and non-federal lands. It ran the department's largest aviation program from the National Interagency Fire Center in Boise, Idaho, flying

The Bureau of Land Management, an agency of the US Interior Department, acquired fifty-seven military-surplus RQ-11 Ravens in the mid-2000s. It also inherited the US Army's fleet of forty-four Honeywell T-Hawks. The hand-launched AeroVironment Raven is shown on display at a trade show exhibit. *Author photo*.

wildfire suppression and monitoring missions, wild horse and burro aerial surveys, and other operations. In the mid-2000s, the BLM acquired fifty-seven military surplus RQ-11A and twenty to thirty RQ-11B Raven hand-launched airplanes, according to Lance Brady, a fire management specialist who led the bureau's UAS program. It acquired the US Army's whole fleet of forty-four RQ-16 T-Hawks. Manufactured by Honeywell, the T-Hawk was a noisy contraption about the size of a wastepaper basket powered by a ducted fan—a propeller mounted within a duct. A memorandum of agreement between the BLM and FAA allowed "broad flexibility of missions that we're able to fly—scientific, wildlife, search and rescue missions," Brady told the 2014 Unmanned Systems Conference. "Basically all Class G (uncontrolled) airspace outside of five nautical miles of an airport, over non-populated lands, at 400 foot AGL (above ground level) or below we're allowed to operate. We have to have specific either written or verbal approvals to fly over other than Department of Interior lands." The memorandum also allowed the BLM to operate at night, he said. A sister science agency, the US Geological Survey, used the Raven and T-Hawk for "persistent monitoring of earth surface processes in areas that have been difficult or nearly impossible to obtain information before."

In 2005, the US Forest Service, a branch of the Department of Agriculture, acquired a CyberBug—essentially a propeller-driven fuselage suspended from a hang-glider wing—to monitor fires. It collaborated with NASA to use the Altair fitted with an instrument pod to map wildfires in California in 2006. The service's Law Enforcement and Investigations branch purchased two hand-launched SkySeer airplanes in 2007, to search for illicit marijuana patches. Those went unused when that year the FAA tightened its regulation of UAVs, according to information the Public Employees for Environmental Responsibility obtained under the Freedom of Information Act.[8] On its website, the forest service said it was "highly interested in new technologies and believes there is potential to use unmanned aircraft systems to support a host of natural resource management activities, including forest health protection, wildfire suppression, research, recreational impacts and law enforcement."

Federal agencies charged with protecting the nation's borders became involved with unmanned aircraft in the 1990s through "cooperative use" of Army Hunter and Navy and Marine Corps' Pioneer UAVs, according to the Department of Defense. The former US Border Patrol sponsored an evaluation of four types of small unmanned aircraft near Laredo, Texas, in April 1999, but determined they were unsuited to its needs.[9] The US Coast Guard started a series of demonstrations in 1999, involving aircraft ranging from the Aerosonde small UAS to the 7,000-pound Altair. As part of its "Deepwater" fleet recapitalization program hatched in the late 1990s, the service intended to buy sixty-nine Bell Eagle Eye unmanned tiltrotor aircraft to operate from its new National Security Cutters, a plan it scrapped in 2007 over cost. The Coast Guard participated with the Navy in a demonstration of the Predator at King Salmon, Alaska, in November 2003, and with NASA in a demonstration of the Altair the following August.[10] Another model the Coast Guard evaluated was the ScanEagle. In May 2013, the service conducted its first drug bust with an "embarked" UAS, flying a ScanEagle from the deck of the National Security Cutter *Bertholf* to help foil a cocaine smuggling operation off the western coast of Central America.[11]

UAVs in Law Enforcement

The security superagency President George W. Bush advanced in the aftermath of the September 11 terrorist attacks quickly pressed drones into domestic service. Congress passed legislation that Bush signed in November 2002 to establish the Department of Homeland Security, a cabinet-level organization crafted of twenty-two different agencies, including the Coast Guard, that officially opened its doors in March 2003. That same year, the DHS evaluated a Predator B scheduled for delivery to the Air Force during an exercise called Operation Safeguard, monitoring illegal crossings along the US-Mexico border near Gila Bend, Arizona. In June 2004, it announced "the first sustained civilian use of UAVs" as part of the Arizona Border Control Initiative to stem illegal border activity, not with the Predator B, but with two Hermes 450s the department leased from Israel's Elbit Systems. Its first procurement contract went to General Atomics in September 2005, for the delivery of one Predator B (serial number BP-101) to patrol the southwest border.[12] The DHS started surveillance of the Canadian border using unmanned aircraft in 2009.

Under the DHS amalgamation, the aircraft fleet the former US Customs Service operated first shifted to the new Immigration and Customs Enforcement agency (ICE), then in 2004 to its sister agency, Customs and Border Protection. CBP consolidated ships and aircraft under its Office of Air and Marine, which became a uniformed division in January 2006. That April, the air branch lost BP-101 when the Predator B "collided with the terrain" ten nautical miles northwest of Nogales International Airport in Arizona during its 118th flight. The National Transportation Safety Board later found that errors by the General Atomics crew that was flying the aircraft at night from Libby Army Airfield in Sierra Vista, Arizona, contributed to the crash. The NTSB's investigation "revealed that the CBP was providing a minimal amount of operational oversight for the UAS program at the time of the accident."[13]

Safety board reports describe other mishaps involving the Predator B, which the military designated MQ-9. According to an NTSB report, the FAA suspended the CBP's certificate of authorization to conduct training at Sierra Vista in November 2008, after an MQ-9 was damaged during a hard landing. The pilot, who at the time was undergoing a checkride, mishandled the landing upon switching from the nose camera view to the view provided by the aircraft's payload, a multi-spectral targeting ball. The FAA reinstated the training COA that December, and the CBP resumed training on February 17, 2009. Two days later, another pilot undergoing training damaged an MQ-9 in a hard landing. In May 2013, an MQ-9 bounced on landing and departed the runway at Cape Canaveral, Florida, sustaining substantial damage.[14] In January 2014, the Office of Air and Marine temporarily grounded the fleet after CBP-159—a Guardian maritime version of the Predator B—experienced a generator failure; prevented from returning the aircraft to Sierra Vista, the crew ditched it in the Pacific Ocean off San Diego.

The CBP's task was to monitor 7,000 miles of shared border with Canada and Mexico and 2,000 miles of coastal water off California and Florida. The Office of Air and Marine claimed to own the world's largest aviation and maritime law enforcement fleet, bringing to bear some 250 airplanes, helicopters, and unmanned aircraft, as well as a tethered aerostat radar system that it inherited from the Air Force, to assist ground units and other agencies combating drug and contraband smuggling, illegal border crossings, and terrorist acts. It assisted the Coast Guard along the nation's maritime borders. It helped agencies such as NOAA and the Federal Emergency Management Agency assess damage natural disasters caused. It supported criminal investigations by ICE, the Federal Bureau of Investigation, the Drug Enforcement Administration, and other federal law enforcement agencies. It even provided air support to state and local police in emergency situations. "We are actually a support agency. We don't go out and drum up our own business," a CBP "air interdiction agent" at Grand Forks Air Force Base in North Dakota told reporters. "We help support all the other three-letter agencies in the US with what they need. When some agency wants something, they put in a request for it, and as long as the request is validated, we will go out and actually do that mission. We assist every state, local and federal agency. If the North Dakota highway patrol wanted us to watch an interstate somewhere for some reason, we would go out and support them also. We operate with a lot of different state and local agencies."

It was an MQ-9 from Grand Forks that participated in what was believed to be the first drone-assisted arrests of US citizens on domestic soil in June 2011. In that incident, the Nelson County, North Dakota, sheriff sought assistance from the CBP to surveil the Brossart family farm, where family members had earlier chased him away with rifles when he came looking for six missing cows.[15] For the agency's border patrol mission, CBP Predators flew in airspace corridors defined by FAA certificates of authorization—along and within 100 miles of the Canadian border and within twenty-five to sixty miles of the border with Mexico, excluding urban areas. These were not routine sentry missions, said the air interdiction agent, who asked to remain unnamed. "A lot of our flying is intel-driven—going to support somebody," he explained. "We don't just put an aircraft up at altitude and keep it there to burn holes in the sky."

With the loss of CBP-159 in the Pacific Ocean, the agency's fleet of MQ-9s stood at ten in September 2014: five "land mission" Predator Bs it used for overland surveillance; one Guardian equipped with the Raytheon SeaVue maritime surveillance radar—a variant of the aircraft the CBP co-developed and used with the Coast Guard; and three "multi-role" variants with optional radar and communications payloads, one of which was the precise Northrop Grumman Vehicle and Dismount Exploitation Radar the Army used in Afghanistan (one accident aircraft was being repaired). The CBP based the fleet at air stations in Sierra Vista, Grand Forks, and Corpus Christi, Texas. Centers in Riverside, California, and Jacksonville, Florida, were capable of controlling missions.

The station at Grand Forks Air Force Base served as the training facility for all DHS employees who flew unmanned aircraft and operated their sensors. With the CBP's Air and Marine Operations Center at Riverside it also hosted one of two "PED" cells (short for "processing, exploitation, dissemination"), which were facilities where the agency captured, crunched, and distributed the data transmitted by its airborne platforms. According to a mandated "Privacy Impact Assessment" the DHS released in September 2013, the CBP's Office of Intelligence and Investigative Liaison was responsible for the post-flight imagery the Predators, helicopters, and fixed-wing airplanes collected.

A US Customs and Border Protection MQ-9 Guardian on static display at the 2011 Paris Air Show. The Guardian, a maritime variant of the Predator B, is distinguished by its belly-mounted Raytheon SeaVue surveillance radar. The ball under the nose is a Raytheon MTS-B electro-optical/infrared sensor. *Author photo*.

"Subsequently, and only upon request, OIIL provides access to the forensic analysis of a particular image and area to authorized persons who have a 'need to know'; when the dissemination is in response to a particular law enforcement activity or case, that analysis may include PII (personally identifiable information)," stated the assessment. The CBP supplied images and data to "identified users, analysts and decision makers" through its "Big Pipe" secure network, which archived selected mission data on a server hard drive for a maximum of seven days.[16]

It would be several years before the secretive FBI acknowledged using drones in domestic skies. The admission came during testimony by its director, Robert Mueller, before the Senate Judiciary Committee on June 19, 2013. "Does the FBI own or currently use drones and if so, for what purpose?" Iowa Republican Senator Charles Grassley asked Mueller matter-of-factly. "Yes, and for surveillance," said the silver-haired lawman, who was then months away from retiring from the post. Grassley asked if the FBI had agreements with federal agencies such as the DOD or the DHS to assist it with drones. Mueller said he was not certain, but did not think so. To the extent there was interagency communication, he added, it would relate to using the airspace. "I think I can assume, since you do use drones, that the FBI has developed a valid set of policies, procedures and operational limits on the use of drones and whether or not (there is) any privacy impact on American citizens," the senator pressed. "We are in the initial stages of doing that and I will tell you that our footprint is very small, we have very few and have limited use, and we're exploring not only the use but also the necessary guidelines for that use," said Mueller. He emphasized that the FBI used drones "in a very, very minimal way, and very seldom."[17]

The following day, Sen. Rand Paul wrote the FBI, seeking more information on its use of drones. When he did not receive a timely response from the bureau, Paul threatened to oppose James Comey—President Obama's nominee to succeed Mueller as FBI director—similar to the way he challenged John O. Brennan as CIA director. The FBI responded in July with classified and unclassified letters. In the unclassified letter, the bureau said it had conducted surveillance using unarmed drones in eight criminal and two national security cases since late 2006. In one case, drone surveillance played a role in the rescue of an eight-year-old boy who was held hostage in an underground bunker in Alabama by retired trucker Jimmy Lee Dykes. The bureau declared it used drones consistent with its own rules and regulations for aerial surveillance, in conformity with Privacy Act restrictions against sharing personal information and with the Fourth Amendment's protection against "unreasonable searches and seizures," and under FAA certificate of authorization. It would secure a warrant in a situation where an individual had a reasonable expectation of privacy under the Fourth Amendment, but to date had not seen reason to do so.[18]

Within a week of Mueller's appearance before the Senate Judiciary Committee, the nonprofit organization Citizens for Responsibility and Ethics in Washington (CREW) filed a Freedom of Information Act request with the FBI, seeking documents "showing the source of these domestic drones, how the FBI paid for them, who trained the FBI to use them, and any policies about this use—including legal justifications." When the FBI declined to expedite its request for information, CREW filed suit in federal court against the Department of Justice—the FBI's parent organization—to compel it to produce documents, and the court later ruled in CREW's favor.[19] That winter, the bureau delivered a first batch of documents that, while heavily redacted, made clear "the FBI's domestic drone surveillance program is on the course of expansion." One document the non-profit organization posted on its website—a white paper—revealed the bureau initially experimented with "small, remotely controlled aircraft" and found them inadequate for video surveillance. Then, in 2005, it evaluated "battle-proven, military UAVs" and determined they "might evolve to meet FBI mission requirements." In 2009, the FBI's UAV inventory and missions "began to increase as the capabilities became known throughout the field." A limiting factor, the paper noted, was the FAA's regulation of UAVs and its concern over their inability to "see and avoid" other air traffic. To address that limitation, the FBI had initiated a joint effort with the FAA, the Department of Justice, and other federal law enforcement entities "to establish a nationwide, standing certificate of authorization for federal law enforcement use, which will greatly expand the FBI's potential deployment scenarios for UAVs."[20]

In September 2013, the Department of Justice inspector general's office released an "interim audit report" on the department's use and support of unmanned aircraft systems. It found that in the nine years from 2004 through that May, the FBI had spent more than $3 million on small UAS and training; the Bureau of Alcohol, Tobacco, and Firearms nearly $600,000; and the US Marshalls Service $75,000. The Drug Enforcement Administration had acquired systems from another federal agency at no cost. Only the FBI and ATF programs were active at the time of the report. The department's Office of Justice Programs (OJP) and the Office of Community Oriented Policing (COPS) had provided $1.2 million in funding to seven local police agencies and non-profit organizations to study or buy unmanned aircraft.[21]

Local Police Adopt Technology

Local police departments showed interest in using unmanned aircraft in the mid-2000s. With the conflicts in Iraq and Afghanistan drawing away state-controlled National Guard units and their helicopters, police and public safety agencies in rural areas were deprived of the air support they counted on for counterdrug operations, searches, natural disasters, and other missions. "A lot of the missions that got flown for these rural law enforcement agencies were not flown because the National Guard helicopters were deployed, pilots were deployed, the state helicopters which maybe had the capacity to support rural law enforcement were too busy doing all the other missions," said Tim Adelman, a pilot and aviation lawyer based in Annapolis, Maryland. "At the end of the day, the rural guys were left without aviation assets."

The National Institute of Justice—the research and development arm of the Justice Department and part of the OJP—assembled advisory groups consisting of "practitioners" from federal, state, local, and tribal agencies that assessed criminal justice technology needs—in areas such as body armor, weapons detection, and aviation—and guided research and evaluation of solutions. The NIJ's assistance was sought to help identify ways in which small, rural police departments could use aviation, said Adelman, who served on the aviation technology working group. "The thought process was that there would be this continuum of cost benefit, with unmanned aircraft at the far left of the spectrum and being relatively low cost, going all the way up to a Bell helicopter at the far right," he said. "That's where we started to see law enforcement start to focus on unmanned aircraft systems."

The NIJ entered into a cooperative agreement with the Sheriffs' Association of Texas in 2006 that included $405,000 in UAS-related funding, according to the Justice

Department inspector general. The Bureau of Justice Assistance—another component of the OJP—awarded grants to the North Little Rock, Arkansas, and the Miami-Dade police to purchase unmanned aircraft in 2007. That year, the COPS office awarded the Gadsden, Alabama, Police Department a grant that included money for UAS. In October 2007, the NIJ issued a technical bulletin informing police of "considerations prior to the purchase and/or operations of unmanned aircraft systems." While UAS held "exceptional appeal to law enforcement," the FAA regulated their use and required a public agency to obtain a certificate of authorization to operate them. A vendor offering to demonstrate an unmanned aircraft system to police needed an FAA-issued experimental airworthiness certificate beforehand, the bulletin advised.[22] The FAA itself was engaged with local police; it sponsored demonstrations by the Houston Police Department using the Insitu ScanEagle in 2007, and by the Miami-Dade Police Department using the Honeywell T-Hawk in 2008.

The early operational experience with UAVs proved inauspicious for some local police. In 2006, the Gaston County, North Carolina, Police Department spent nearly $30,000 on a CyberBug, ground control station, operator training package, and infrared camera, according to an invoice the news site MuckRock obtained through a Freedom of Information Act filing.[23] "Gaston County police had tested the device on a brisk day in February 2006. But with cameras rolling and reporters watching, the 5-foot unmanned aircraft broke a rudder and wouldn't turn left. Then it had trouble picking up a signal from the global positioning satellites that help guide it, even though it was being controlled by employees for the company that made it," a Charlotte television station reported years later, when drones became controversial. The department later shelved the CyberBug it had acquired with "drug asset forfeiture funds."[24] The COPS grant to the Gadsden police, which included $150,000 for UAS-related costs, focused on stopping the spread of methamphetamine. In 2009, the department deployed one of its two AeroVironment Wasps on a surveillance mission. the Justice Department's IG office reported:

> [D]uring the mission the ground control station lost communication with the UAS, causing the UAS to collide with a tree. Officials told us that they believe that the rugged topography (hills and valleys) of the mission area led to the communication problem. According to the FAA, the Gadsden Police Department did not obtain the required COA prior to its mission. The Gadsden Police Department reports that it has kept the UAS in a secured storage facility since that time, and that it has yet to determine whether or how it will use its UAS in the future.

In its audit report, the IG was critical of the Justice Department's grantmaking process. "[N]either OJP nor COPS required that UAS award recipients demonstrate that they could receive FAA approval to operate UAS or that UAS use was legal in their jurisdiction," the report stated. "Their UAS awards also did not require recipients to report specific data necessary to measure the success of UAS testing, or to use or share the results of their programs with DOJ." While local police use of UAS could interfere with the department's own surveillance efforts, "we found no evidence that OJP or COPS coordinated with or notified DOJ law enforcement components about their UAS awards, either before or after the awards were made. Moreover, we note that OJP had difficulty confirming to us how many UAS-related awards it had made." The IG recommended the offices tighten their tracking of UAS awards, require applicants to demonstrate an ability to obtain FAA authorization, establish reporting requirements to measure the effectiveness of UAS awards, and notify other branches of future awards.

A program the Federal Emergency Management Agency started in 2004, which later shifted to the science and technology directorate of the DHS—its parent organization—assisted emergency response agencies in deciding which equipment they should buy. The System Assessment and Validation for Emergency Responders (SAVER) program tested and evaluated commercially available products falling within twenty-one categories of a DHS authorized equipment list. Examples of categories were "search and rescue," "surveillance," and "chemical, biological, radiological, nuclear and explosive detection." Within these were authorized items for which the program provided equipment information documents. "In any given year, the SAVER program evaluates between 30 and 40 products," stated an article the NIJ published in 2011. "Since its inception, the program has evaluated more than 400 products and published more than 800 equipment information documents. A few examples of the myriad types of equipment for which information is available include in-car cameras, law enforcement protective helmets, small bomb disposal robots, thermal imaging cameras, hydraulic rescue tools and facial recognition technology"[25] Other authorized products included remotely controlled ground vehicles, submersible vehicles, and unmanned aircraft. More than twenty grant programs supported equipment purchases from the DHS list.[26]

The fiscal year 2009 National Defense Authorization Act that President George W. Bush signed in October 2008 contained a "sense of Congress" resolution calling on the FAA and the DOD to form a joint executive committee "on conflict and dispute resolution" focused on unmanned aircraft. As Congress saw it, the committee of senior leaders would broker disagreements over policy and procedure arising from the Pentagon's need to routinely access the nation's airspace system for UAS training purposes and to support its regional commands and "domestic authorities in emergencies and national disasters." In March 2009, the US Northern Command hosted a "UAS Summit" in Colorado Springs, Colorado, which was described in minutes from the event as the first such meeting of senior leaders from federal agencies with a stake in using the airspace for unmanned aircraft, including the DHS, DOE, and the Department of Interior. "Throughout the two-day summit, numerous issues were discussed with 'opportunities' highlighted for each organization and agency to assist in making less restrictive UAS-NAS (national airspace system) integration a reality," stated a copy of the minutes a source provided the author. The summit concluded that "UAS-NAS integration issues must be solved using the entire 'federal family' and should not be tackled by individual organizations and stovepipes. The appropriate experts should tackle the issues at the right place and time, providing the necessary feedback to the entire community of participants."

In late 2009, the FAA and the DOD formed the UAS Executive Committee the Congress desired and invited two other major players—NASA and the DHS—to participate. Absent from the group was the Justice Department and law enforcement, among the biggest potential users of unmanned aircraft, observed Adelman. "Within the FAA, unmanned aircraft systems were very much being driven by DOD and law enforcement didn't have much of a voice at the table," he said. The NIJ's aviation technology working group tried to interest the FAA but wasn't initially successful. "We were talking but the FAA wasn't responding," he said. Around 2010, FAA aviation safety inspector and former US Marshals Service pilot David Morton and former FAA and Transportation Security Administration executive Steven Pansky supported involving law enforcement agencies—and that was a turning point, according to Adelman. He also credited the office of California Republican congressman Howard "Buck" McKeon for bringing attention to police use of unmanned aircraft.

In 2012, the DHS science and technology directorate launched a new program called RAPS (Robotic Aircraft for Public Safety). Under RAPS, the department tested small drones and produced reports on their performance for consumption by local police, disaster response, and fire fighting agencies considering using unmanned aircraft in their operations. The program chose as a test site an Oklahoma State University affiliated facility near Elgin, Oklahoma, that had access to restricted airspace above Fort Sill, a US Army base. "Law enforcement, unfortunately, can be the victim of salesmen," said Adelman. "Someone comes in and has a nice shiny toy and says it will do X, Y and Z and it will solve all your law enforcement problems. We have to flip the table on that. We have to say—law enforcement you need to identify your problem, set your criteria to solve that problem and then go find the technology. I think that's what RAPS will ultimately help do." Speaking to a meeting of AUVSI's Oklahoma chapter in March 2013, Kirk Kloeppel, who worked for a DHS contractor on the RAPS program, said seventy companies responded to the government's request for information to participate in the flight demonstrations. Twenty were eliminated from consideration because their aircraft were either too heavy or not adequately developed. The program required that UAS types weigh less than twenty-five pounds and have at least 1,000 flight hours, said Kloeppel, a retired colonel who previously served as the director of the Air Force Research Laboratory's munitions directorate.[27] As of June 2014, it had tested seven small airplanes and eleven helicopters from seventeen companies.[28]

When it came to police use of unmanned aircraft, presenters often noted that only about 200 of the 18,000 state and local law enforcement agencies in the United States operated an aviation unit to begin with, according to Bureau of Justice Statistics, but the comparison was uneven. When counting just larger agencies of 100 or more sworn officers, the BJS found that one in five had an aviation unit in 2007, with 899 total aircraft. Half of the 201 aviation units operated only helicopters, representing two-thirds of the national fleet, but none of the units reported operating unmanned aircraft.[29] Seven years later, subject matter experts estimated that a dozen or fewer local police agencies actively used drones. The examples that stood out were smaller agencies with jurisdiction over large tracts of land with lesser population density than in urban areas. Those agencies cited the cost of owning and operating aircraft—an absolute value no matter the jurisdiction—as motivation for using unmanned machines.

The Mesa County, Colorado, Sheriff's Office was featured in *Time* magazine's "Rise of the Drones" cover story. Based in Grand Junction—the county seat—the office employed 200 people, of which sixty-five were deputies. Its area of responsibility was a county with 175,000 residents covering 3,300 square miles. It reported first testing an unmanned aircraft in October 2008. The following year, the agency acquired a Draganflyer X6 helicopter the manufacturer donated to demonstrate its benefits for law enforcement. It flew the three-armed, six-rotor machine over the county landfill under FAA certificate of authorization. In 2010, the FAA modified the COA to allow the agency to deploy the Draganflyer county-wide. In January 2012, the sheriff acquired a bungee-launched Falcon UAV airplane whose manufacturer "ultimately only charged us for the materials to make the system."[30]

The cost of operating the unmanned models was $25 per hour—a fraction of the $250 or more the office would pay for a manned aircraft, said Ben Miller, the sheriff's unmanned aircraft program manager. "It may surprise you (that) we don't do this really from a capability perspective. We do what we do because we operate (with) a budget of about $25 million, and most of that is staff compensation and inside that budget we operate a jail facility of 336 beds. We're not a large agency that can afford manned aviation," Miller told the Unmanned Systems Conference in 2014. Later that year, the Mesa County public works department added a third model to the county's fleet—Trimble's UX5 flying wing—which it used for aerial surveys of the landfill and road and bridge construction projects.

The Grand Forks County Sheriff's Department in North Dakota told a similar story. Its unmanned aircraft program was a research project that started in 2010 with a grant from the University of North Dakota and contributions from UAV manufacturers AeroVironment and Draganfly Innovations. The project's objective was to develop a "concept of operations" for police missions using small unmanned aircraft and a training curriculum for pilots and sensor system operators. The university leased four models from the manufacturers—AeroVironment's Qube quadcopter and Raven hand-launched airplane and Draganfly's X6 and X4ES helicopters—which it sub-let to the sheriff's department. The department practiced with unmanned aircraft for eighteen months under certificates of authorization the FAA required for training before securing an operating COA in November 2012. The first COA allowed it to fly the Draganflyer X6 within sixteen counties in northeast North Dakota that belonged to the same emergency services mutual aid region. In March 2014, the department received an FAA operational COA to fly at night, the first police agency in the country so authorized. Six months

The Grand Forks, North Dakota, Sheriff's Department used the AeroVironment Qube quadcopter (shown here) and the Draganfly Innovations' Draganflyer X4-ES quadcopter in its operations. In September 2014, it used the Qube at night to help apprehend suspects, a US law enforcement first. *Author photo.*

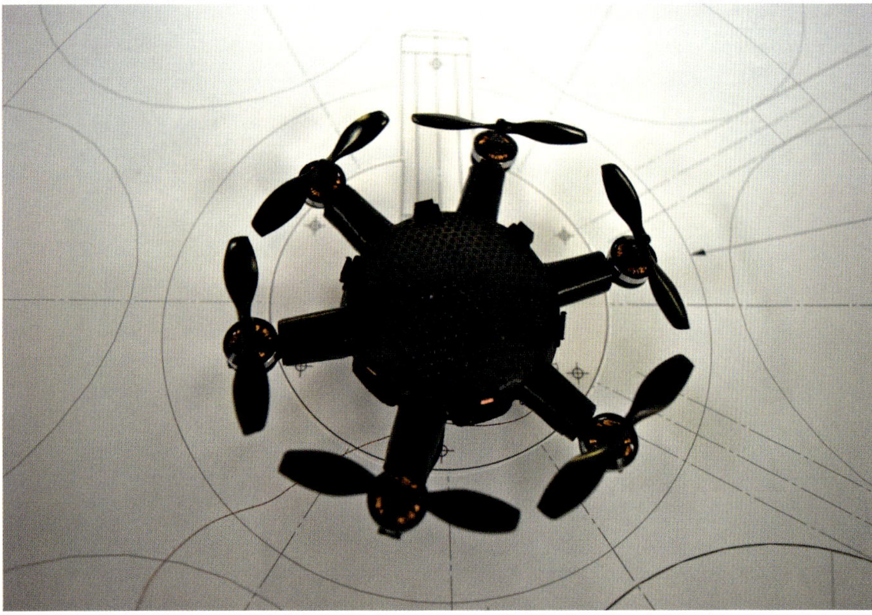

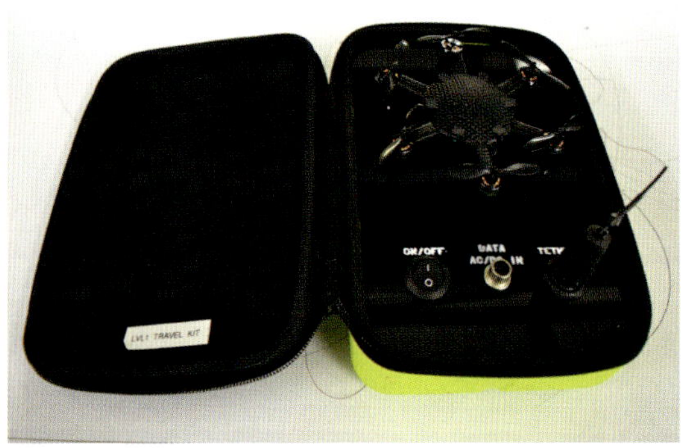

CyPhyWorks of Danvers, Massachusetts, supplied the three-ounce Pocket Flyer (top), which could operate indoors or in environments where GPS or radio frequency signals were denied. It was powered by a microfilament tether that spooled from inside the aircraft so if the tether became snagged anywhere, the aircraft kept moving. The tether supported power to the aircraft and command and control by an operator and HD video. The Pocket Flyer carrying case (right) fit into a cargo pants pocket. *Author photos*

later, it deployed the Qube at night to apprehend drunken driving suspects hiding in a cornfield—another first.

Advising the project was Alan Frazier, an associate professor with UND's aviation department and a former policeman and aviation unit supervisor with the Los Angeles suburb of Glendale, California—in his words "a cop and pilot who sort of stumbled into academia." The Glendale police owned four turbine helicopters worth $10 million when he left in 2005, with a budget of $1.8 million to fly and maintain them. "We can run an unmanned aircraft unit for $10,000 a year," he said of the Grand Forks department. "We've done that easily—paying for operators, paying for their training, paying for the insurance on the aircraft, the trickle of electricity that recharges our lithium polymer batteries. It's the difference between almost $12 million a year to run a medium-sized air support unit or $10,000 a year to run a unit that has four aircraft." Obviously, four full-scale helicopters were more capable than a fleet of miniature flying machines, he allowed. "We inserted SWAT members, we moved K-9s to remote locations, we did all kinds of things that we can't do with a small unmanned aircraft. But one of the things this research is flushing out for us is that we can do about one-third of the missions of a manned air support unit, but we can do it for a fraction of the cost," Frazier said.

Nevertheless, what Adelman described as a "cultural issue" among agencies that had never used aircraft of any kind in their work, local government funding constraints, and mounting concern over the privacy implications of drones helped slow their proliferation to police departments on the home front. "There's a difference between having a certificate of authorization to fly one and actually using them," Adelman said in the fall of 2014. "There are not many agencies out there that are actually using them effectively and getting real functionality out of them right now. We're still very much at the tip of the iceberg when we talk about law enforcement use."

CHAPTER 4

THE FAA MODERNIZATION AND REFORM ACT

President Barack Obama signed the FAA Modernization and Reform Act into law on February 14, 2012—Valentine's Day. But it wasn't love or even affection that produced the legislation. Democrats and Republicans in the US Congress bickered with each other for more than four years before passing the FAA reauthorization bill—the term used for legislation that reestablishes funded programs within an agency. Lawmakers clashed over a labor provision affecting airline workers' ability to organize, over subsidized flights to rural airports, and over the number of flights servicing Ronald Reagan Washington National Airport across the Potomac River in Northern Virginia—the nearest airport for politicians commuting to the nation's capital. The dysfunction affected more than just aviation. Congress had not renewed the surface transportation bill since it expired in 2009, nor was the climate conducive to companionship. The 2010 midterm elections returned control of the House of Representatives to a Republican Party infused with recalcitrant "tea party" Republicans committed to gutting federal spending and stopping the President's signature health-care reform legislation that had passed earlier that year.

Congress had passed twenty-three short-term funding extensions since the last FAA reauthorization expired in September 2007. Disagreement between the Democrat-controlled Senate and the Republican-controlled House over one such extension partially cut off funding to the FAA between July 23 and August 5, 2011. At issue: the $200 million Essential Air Service, a federal program that supported airline service to little-used, mainly rural airports. The House sought to severely curtail the program in its version of the FAA funding extension; the Senate disagreed. The stalemate caused the FAA to furlough 4,000 employees and stop work on 219 airport construction projects. More than $300 million in federal taxes on airline tickets went uncollected. It ended when Transportation Secretary Ray LaHood announced he would issue waivers for some of the airports the House wanted eliminated, leading the Senate to accept the House bill.

As the divided Congress reeled toward its next self-imposed deadline to extend the FAA's operating authority, representatives of airline pilots' and flight attendants' unions staged a press conference at Reagan National airport to warn of the potential safety

Air Line Pilots Association president Lee Moak spoke during a press conference unions held at Washington Reagan National Airport in September 2011, urging Congress to pass long-term FAA reauthorization legislation. Veda Shook, then president of the Association of Flight Attendants-CWA, also participated. *Author photo*.

consequences of unstable funding to the nation's aviation infrastructure. "We're in critical need of a real bill, a bill that will provide the necessary level of funding to enhance the future safety and efficiency of our aviation system.…To do anything less at this moment jeopardizes safety and has a direct impact on the viability of the entire industry," said Air Line Pilots Association president Lee Moak, a Delta Air Lines captain.[1]

Aviation industry executives also fretted over the damage done by federal funding instability to the FAA's ambitious future vision for air traffic control modernization known as the Next Generation Air Transportation System (NextGen). Clayton Jones, then chief executive of avionics manufacturer Rockwell Collins, posed a troubling question at the RTCA Symposium held in Washington, DC, in June 2012. "Who in this room," Jones inquired, "has any reasonable degree of confidence we're going to actually get the funds necessary to implement NextGen by 2020?" Without adopting an approach to bankrolling NextGen based on public-private partnerships, he said, "we are doomed to hear the same luncheon keynote speech like this one every year for the next 10 years."[2]

The mainstream media often described NextGen as the transition from a radar-based to a satellite-based air traffic control system, but it was much more than that. NextGen promised a system in which airplanes navigate the airspace by precise GPS position—yes, but also one in which pilots and controllers communicate by text messages instead of by voice and airliners fly "trajectory-based," or gently ascending and descending flight paths with no fuel-consuming step-downs and holding patterns. Airports easily manage thousands of daily arrivals and departures through complex, computerized sequencing. Unmanned aircraft systems also figured in that vision. "With the introduction of the sUAS Rule for civil operators, there will be an increase in the number and scope of UAS flights in an already busy" national airspace system, the FAA advised in a 2011 fact sheet.[3] "Decisions being made about UAS airworthiness and operational requirements must fully address safety implications of UAS flying in the same airspace as manned aircraft, and perhaps more importantly, aircraft with passengers. Overcoming these challenges associated with the differences between manned and unmanned aircraft while simultaneously transitioning to NextGen further amplifies the need for extensive cooperation between the FAA, other government agencies, and industry." Congress laid the foundation for NextGen in a 2003 legislation called the "Vision 100—Century of Aviation Reauthorization Act." The FAA was spending about $1 billion annually on development programs; NextGen's estimated cost to government and industry over two decades lay somewhere north of $40 billion, assuming all of the program elements fell into place.

In late January 2012, House and Senate negotiators finally reached compromise on a four-year, $63 billion reauthorization bill to fund the FAA through fiscal year 2015. The full House approved the measure (H.R. 658) on February 3. The Senate endorsed the legislation three days later, sending it to the president for his signature, and with that it became Public Law 112-95. "For four years, Democrats in control of Congress left the nation's aviation system in the lurch by failing to pass a long-term bill to reform FAA programs and set national aviation policy and priorities," declared Florida Republican John Mica, the House Transportation and Infrastructure Committee chairman who introduced the legislation. "Today we have in place sound multi-year policies that reform FAA programs, eliminate expensive ticket subsidies, modernize our air traffic control system, improve airport infrastructure, reduce air traffic delays and create jobs," added the combative Mica, who was complicit in delaying the legislation.[4]

To the aviation industry and the American traveling public at large, the bill's passage brought long-delayed funding stability to the FAA and renewed hope for accomplishing the NextGen vision, yet the political warfare over various provisions distracted from the legislation's profound significance to the unmanned aircraft systems industry. Couched within it were several pages devoted to introducing unmanned aircraft into the national airspace system, laying the groundwork for their transition from military machines deployed in remote war zones to commercial products used in civilian airspace at home. The language did not go entirely unnoticed. "Within a few years, that flying object overhead might not be a bird or a plane, but an unmanned aircraft," reporter Bart Jansen wrote in *USA Today* after the Senate voted to approve the reauthorization bill. "Drones, perhaps best known for their combat mission in Afghanistan, are increasingly looking to share room in US skies with passenger planes." The legislation "would require the Federal Aviation Administration to devise ways for that to happen safely in three years."[5]

Indeed, under Subtitle B the statute called upon the secretary of Transportation, in consultation with the aviation industry, other federal agencies, and the unmanned aircraft systems industry to develop a "comprehensive plan" that provided for the "safe integration of civil unmanned aircraft systems into the national airspace system as soon as practicable, but not later than Sept. 30, 2015." The plan was to define standards for operation and certification of civil unmanned aircraft and ensure that any civil system included a "sense and avoid" function, which the Congress defined as "the capability of an unmanned aircraft to remain a safe distance from and to avoid collisions with other airborne aircraft." The plan would establish UAS pilot and operator registration and licensing standards, and describe a "phased-in approach" to introducing unmanned aircraft into the airspace. Congress instructed the FAA to fold the plan into its annually updated NextGen Implementation Plan document.

Further, lawmakers directed the FAA to produce a five-year, annually updated "roadmap" for introducing unmanned aircraft into the airspace. Not later than eighteen months after delivering the comprehensive plan, the agency was to publish a final rule on small unmanned aircraft. Some six months after the plan's enactment, it was to establish a program to introduce UAS at six national test ranges—a requirement Congress had also inserted in the fiscal 2012 National Defense Authorization Act. Under the heading "Expanding Use of Unmanned Aircraft Systems in the Arctic," the statute required the FAA to "designate permanent areas in the Arctic where small unmanned aircraft may operate 24 hours per day for research and commercial purposes." Within the designated permanent areas, the law instructed the agency to develop procedures for beyond line-of-sight, over-water UAS flights to at least 2,000 feet in altitude.

A separate section addressed public unmanned aircraft. Congress directed the secretary of Transportation to speed the process of issuing certificates of authorization for public machines and to enter into agreements with other government agencies to routinely issue them. The legislation allowed "a government public safety agency" to operate an unmanned aircraft weighing 4.4 pounds or less, as long as it flew the aircraft within line of sight of the operator, less than 400 feet above the ground, during daylight conditions, within Class G (uncontrolled) airspace, and more than five miles from an airport. The FAA was to produce operational and certification requirements for public unmanned aircraft by December 31, 2015.

The 2012 legislation "is probably going to be a watershed event when we look back in time in another 50 years or so," observed AUVSI's Michael Toscano. "What it basically did was it forced from above the integration (of unmanned aircraft) in a specific time. This allowed the game to be turned up on establishing 'when you could,' and it started to identify low-hanging fruit" to introducing unmanned aircraft into the nation's airspace system. The industry and legislation were tightly linked. The self-described anti-corruption blog *Republic Report* caused an Internet stir when it revealed a PowerPoint presentation indicating AUVSI helped write the unmanned aircraft provisions of the FAA reauthorization act.[6] Toscano acknowledged that his organization influenced the legislation—a function that is not unusual for trade associations. "Success has many fathers, failure has but one," he said. "Many people will claim the credit for that (legislation), including us. We were instrumental in providing language to the lawmakers that were doing this. As far as the language, yes, AUVSI helped participate in the crafting of the words and the language that was utilized in the final authorization act." The lawmakers who came to lead the House and Senate UAS caucuses and the Congressional Robotics Caucus deserved most of the credit for spearheading the legislation in Congress, Toscano said. Reps. Mike Doyle—a Democrat from Pennsylvania—and Phil Gingrey—Republican of Georgia—were co-chairmen of the robotics caucus that lawmakers formed in 2007. Reps. Buck McKeon (Republican of California) and Henry Cuellar (Democrat of Texas) chaired the House Unmanned Systems caucus formed in 2009. Senators Jim Inhofe, (Republican of Oklahoma) and Joe Manchin (Democrat of West Virginia) led the Senate UAS caucus, which they announced in September 2012, after Congress passed the FAA reauthorization legislation.

Theodore "Ted" Wierzbanowski, a towering, genial former Air Force colonel and experimental test pilot who went to work for AeroVironment after retiring from military service, said industry groups, including the UNITE alliance, became involved as early as 2005, when Congress was beginning to draft what would eventually become the FMRA. "Through UNITE [and] literally every organization, all the companies tried to somehow influence the language," Wierzbanowski recalled. "It took from probably '05 until '12 because [the Congress] just kept 'kicking the can,' they never passed the bill. It kept changing and more people got involved and more people started to get interested in this. It was fascinating to watch; it's the political process." Major aerospace manufacturers had formed UNITE in mid-2002 "to encourage national action to open the national airspace to high altitude, long endurance UAVs," according to the organization's website. It listed AAI Textron, AeroVironment, BAE Systems, Boeing, General Atomics, Lockheed Martin, Northrop Grumman, Raytheon, and Rockwell Collins as members. The alliance originally supported NASA's Access 5 project to integrate UAVs in the national airspace system. When the space agency cancelled the project in 2006, UNITE remained active but "behind the scenes." As legislation supporting the introduction of civil unmanned aircraft took root, UNITE's directors instructed its management "to increase collaborative activity and to no longer

US Rep. Howard McKeon, a California Republican, was chairman of the House Armed Services Committee and sponsored the National Defense Authorization Act, requiring the FAA establish six unmanned aircraft test ranges. Rep. Mac Thornberry (left) succeeded him as chairman when McKeon retired from Congress in 2014. *Author photo.*

work 'behind the scenes.' This has resulted in what has been 'perceived' as a reemergence of UNITE when, in reality, it is just that UNITE has made a conscious decision to become more visible and more active."

An investigative article by the Center for Responsive Politics and Hearst Newspapers that was published in the *San Francisco Chronicle* and the *Houston Chronicle* in November 2012 found that members of the fifty-strong House Unmanned Systems Caucus had helped push the aerospace manufacturers' "agenda" and drawn "nearly $8 million in drone-related campaign contributions" in the previous four years. Representatives from California, Texas, Virginia, and New York received most of the funds. McKeon, who was chairman of the House Armed Services Committee and a staunch supporter of the military, led the pack, collecting $833,650 in "drone related" campaign contributions. He represented California's 25th district, including the city of Palmdale, where Northrop Grumman assembled the RQ-4 Global Hawk. McKeon retired from Congress in 2014, after serving twenty-one years in the House; the press office of the armed services committee declined to make him available for an interview. In total, eleven "drone caucus lawmakers" from California received $2.4 million from manufacturers' political action committees and employees during the 2010 and 2012 election cycles, according to the Center for Responsive Politics' tabulation of Federal Election Commission reports. Among the largest donors to caucus members were Boeing, Honeywell, Lockheed Martin, Raytheon, and General Atomics.[7]

The Northrop Grumman RQ-4 Global Hawk, shown here at Grand Forks Air Force Base in North Dakota, served the US Air Force as a high-altitude, long-endurance surveillance platform. It had a maximum gross takeoff weight of 32,250 pounds and a 130-foot wingspan, exceeding that of the Boeing 737 airliner. Northrop Grumman manufactured fuselage sections in Moss Point, Mississippi, then assembled the aircraft in Palmdale, California. *Author photo.*

The reauthorization legislation may have indeed "forced from above" mandates the unmanned aircraft industry sought, as Toscano said, but the FAA started slipping immediately on the various deadlines Congress established. In a September 2012 report to Congress, the Government Accountability Office listed one accomplished deadline to that point—the FAA had expedited its process of issuing COAs to public safety agencies to fly unmanned aircraft, which it now granted for a term of two years instead of one. But the agency lagged in most other areas, including establishing the test-range program, developing an Arctic operations plan, and determining whether certain unmanned aircraft could already safely fly in the national airspace system under the Section 333 provision—steps Congress wanted completed by that summer.[8] The FAA expected that its small UAS proposed regulation would be published in the *Federal Register* for public consumption by late 2011, but the small UAS notice of proposed rulemaking was nowhere to be found. The website *sUAS News* featured a ticker it named "FAA Miss-O-Matic" to measure the time to the minute since the draft rule was due.

Marking progress toward one of the requirements of the 2012 act—to designate Arctic operational areas for UAS—the FAA announced in July 2013 it had issued restricted category type certificates for the Insitu ScanEagle and AeroVironment Puma, for the first time permitting operators to fly those aircraft for commercial purposes. On September 12, 2013, ConocoPhillips and Insitu conducted the first FAA approved commercial unmanned aircraft flight, launching a ScanEagle from the *Westward Wind* in the Chukchi Sea.

Some nine months later than Congress specified, the FAA unveiled its first five-year "Integration of Civil Unmanned Aircraft Systems in the National Airspace System Roadmap" on November 7, 2013, at a "Civil UAS Forum" the Aerospace Industries Association convened on short notice at a downtown Washington, DC, hotel. Earlier in the year, the association had upgraded its UAS subcommittee to a

Marion Blakey, president and CEO of the Aerospace Industries Association, previously served as chairman of the National Transportation Safety Board and administrator of the National Highway Traffic Safety Administration. She was FAA administrator when Congress passed legislation launching the NextGen modernization effort in 2003. *Author photo.*

full committee, anchored by manufacturers Boeing, Lockheed Martin, and Northrop Grumman. It had organized a panel discussion at the Paris Air Show that June to discuss the findings of a poll it conducted with the *Christian Science Monitor* to gage public perceptions of the non-military use of unmanned aircraft. AIA, then chaired by Northrop Grumman CEO Wes Bush, was stepping up its profile on civil unmanned aircraft.

AIA president and CEO Marion Blakey, who in an extraordinary career had served terms as administrator of the FAA, chairman of the National Transportation Safety Board, and administrator of the National Highway Traffic Safety Administration, gave the opening remarks. "Welcome to an event which represents a very important milestone for the integration of unmanned aircraft systems into our national airspace," she told the assemblage of reporters, FAA staffers, and association employees. Unmanned aircraft represented America's next aviation frontier, said Blakey, whose soothing Alabama accent remained strong despite years in Washington. "Our association, which represents a number of leading UAS manufacturers, has made it a priority to champion UAS and their many beneficial civil applications. You've heard a lot about these systems in various ways, sometimes very distorted and negative reports. We think it's critical to not lose sight of the variety, in fact the enormity, of the benefits that await our society with this breakthrough technology. In short, UAS promise to improve lives in a number of ways."

Blakey then introduced FAA Administrator Michael Huerta, who used similarly soaring words to capture the moment: "NextGen is...allowing us to create the airspace of the future, and in that airspace of the future we will have new users. We will have more commercial space launches and we will have more unmanned aircraft systems," Huerta advised. "This document, which has been developed with key stakeholders, outlines what we need to do to safely integrate unmanned aircraft into our national airspace."

In the seventy-page inaugural road map, the FAA set out its vision for introducing unmanned aircraft in three stages: "accommodation," "integration," and "evolution." Accommodation was the status *du jour* and would prevail during the near-term period of less than five years, the agency said. During this time, it would consider applications to operate UAS in the airspace system on a case-by-case basis. Companies could gain entry for their aircraft the traditional way, by obtaining a special airworthiness certificate in the experimental category. Research and development would continue to address technological barriers and "validate advanced mitigation strategies" to ensure safety. The integration stage would begin in five to ten years, with the implementation of the small UAS regulation. During this stage, the FAA would set "threshold performance requirements" for unmanned aircraft and produce new or revised rules, policies, procedures, guidance material, and training to support "routine" airspace operations. All supporting materials would be in place and routinely updated during the dynamic evolution stage, which the FAA expected to happen in the long-term period 2022 to 2026.[9]

Simultaneously with the road map's release, Huerta unveiled an FAA-developed privacy policy the agency planned to implement at the still-unspecified UAS test sites Congress mandated in the reauthorization act. "This policy requires operators to comply with all local, state and federal laws concerning privacy and civil liberties," he said. "We're requiring test site operators to create a privacy policy and to make it available to the public. And they must require anyone operating unmanned aircraft at the test sites to have a written plan for how they will use and retain any test data that they have acquired." He introduced yet a third document: the congressionally required UAS Comprehensive Plan, prepared by the FAA-led multiagency Joint Planning and Development Office, which would later be disbanded. The comprehensive plan, resulting from the collaboration of the FAA, NASA, and the Departments of Defense, Commerce, and Homeland Security, set "the overarching, interagency goals, objectives and approach to integrating UAS" in the national airspace system and helped coordinate the partner agencies' research and development efforts. It established several "national" goals, among them achieving routine access to the airspace for small public and civil unmanned aircraft flown within line of sight—without a COA or special airworthiness certificate—by 2015. Routine access by larger public machines was set for 2015, and for civil machines in 2020. "Achieving approval of the UAS National Goals and Objectives…was a key accomplishment, since this allowed the stakeholders to work in unison," stated the plan. "[T]he overarching approach for the Goals is to allow public integration to lay the framework for civil integration."[10]

Though late, the FAA would honor another major requirement of the reauthorization act in 2013. For months, the agency had promised in public forums that it would choose the six national UAS test ranges by the end of the year. While there was no federal funding attached to the test site designations, half of the country eagerly sought the FAA's endorsement. State government agencies and research institutions from twenty-four states combined to submit twenty-five different proposals, many representing multiple chunks of airspace. On December 30, the FAA announced it had selected universities and other public entities in Alaska, Nevada, New York, North Dakota, Texas, and Virginia to operate the test ranges. Entities in several other states would participate in the testing activities through partnerships with the lead organizations. Huerta broke the news in a morning teleconference with reporters. "Across the six applicants, we're confident that our research goals of system safety and data gathering, aircraft certification, command and control link issues, control station layout and certification, ground and airborne sense-and-avoid and environmental impacts will be met," he said. "What we have is the platform to conduct broad-based research, considering a wide variety of different factors."[11] Reporters wanted to know how the test site program that Congress approved to run through February 2017 synched with its overarching direction to introduce UAVs into the airspace system by September 30, 2015. "I'm a little confused about the sequence here because the (press) release says that they can keep flying until February of '17," said Matthew Wald of *The New York Times*. "In other words, this research will overlap the introduction of this equipment into the national airspace system; it won't be done by the time that these things are certified, or will certification be delayed?" Huerta responded: "No, it's an ongoing process…As we are conducting this research, I think the best way to think about it is it would be a staged integration of unmanned aircraft into our national airspace system."

Nearly two years after Congress passed the FAA Modernization and Reform Act of 2012, the office of Department of Transportation Inspector General Calvin L. Scovel III identified safely introducing unmanned aircraft into the national airspace system among key challenges the FAA faced. "FAA states that problems in meeting the act's requirements are due to the complexity of the problem, privacy issues and unresolved coordination issues with other federal agencies," the IG stated in its "Top Management Challenges for FY2014" report to Congress. UAS-specific safety issues, such as their ability to "detect, sense and avoid" other air traffic, remained unresolved. Although unmanned aircraft already plied the airspace system, the FAA had not developed air traffic control procedures for "safely co-managing" manned and unmanned aircraft.[11] In written testimony he presented to Congress on February 4, 2014, Gerald Dillingham—the GAO's director of physical infrastructure issues—said the FAA had achieved seven nearer-term requirements of the reauthorization act, but six were still "in process." While the agency had made progress, Dillingham testified, "integrating UAS into the national airspace system continues to challenge FAA, leading to uncertainty about when UAS integration will be achieved."[12]

Congress took heed of the warnings. In a report accompanying their version of the 2015 Transportation, Housing, and Urban Development and Related Agencies Appropriations Act, House appropriators were "concerned that the FAA may not be well positioned to manage effectively the introduction of UAS in the United States." They provided an additional $3 million to the FAA's Aviation Safety budget to expedite introducing unmanned aircraft.[13] Senate appropriators recommended a $4 million bump in research (to $12.9 million), all of it devoted to the FAA's planned UAS center of excellence. They directed the FAA to address "research gaps" by collaborating with other federal agencies and coordinating the efforts of the center of excellence and the six test sites.[14]

Northern Virginia Omni Versatile Solutions demonstrated its N-Cognito at Kennedy Space Center in Florida during the Unmanned Systems 2014 conference. The FAA later granted the firm a Section 333 exemption to fly the 7.7-pound airplane for precision agriculture. *Author photo.*

As the appropriations legislation moved through Congress, Scovel's office released an audit report in the summer of 2014 that said, flatly, the FAA would not make the September 30, 2015, deadline to introduce unmanned aircraft into the nation's airspace system due to unresolved technological, regulatory, and privacy issues. "[W]hile it is certain that FAA will accommodate UAS operations at limited locations, it is uncertain when and if full integration of UAS into the national airspace system will occur," the IG said. As of the report's release, the agency had completed nine of the seventeen UAS-related provisions of the 2012 reauthorization act.[15] By December 2014, the FAA administered 300 active public-use COAs, eighteen experimental airworthiness certificates, and two restricted category tickets for unmanned aircraft. However, despite the case-by-case authorizations, the agency "will not meet the act's goal of achieving safe UAS integration by September 2015," Assistant IG Matthew Hampton reiterated in testimony before the House Aviation Subcommittee.[16]

CHAPTER 5

FAA STANDS UP UAS INTEGRATION OFFICE

Michael Huerta was not supposed to be FAA administrator. When President Obama took office in 2009, he named former airline pilot and labor leader J. Randolph "Randy" Babbitt to head the $15 billion agency and its 47,000 employees. Obama nominated Huerta in December that year to be deputy administrator—the agency's No. 2 man. In that capacity Huerta was to spearhead the ambitious NextGen program to modernize the nation's air traffic control system, which was gathering steam. As Washington would have it, Huerta's appointment was blocked by Republican Senator John Cornyn; the Texas lawmaker wanted the FAA to expedite approval of a certificate of authorization allowing the US Customs and Border Protection agency to operate Predator UAVs from the Corpus Christi Naval Air Station, closing a surveillance gap along the border of Texas and Mexico. The FAA had already authorized the CBP to overfly the border from El Paso to Big Bend National Park in West Texas; the state's congressional delegation wanted the flights extended to the Gulf of Mexico. After meeting with Babbitt, Cornyn announced on June 23, 2010, that the FAA had approved the COA.[1] The Senate confirmed Huerta as deputy administrator the same day.

That a drone would delay Huerta's appointment was fitting. Less than two years after he took over as deputy administrator in charge of NextGen, the phenomenon of unmanned aircraft would be suddenly his to manage, the requirements of the FAA Modernization and Reform Act his to achieve. Babbitt headed the agency as the wave gathered; in a November 2009 speech to the Aerospace Industries Association in Scottsdale, Arizona, he revealed that the number of unmanned aircraft operations the FAA authorized in civilian airspace had more than tripled since 2007, but he tamped down any expectation the agency was close to allowing wider access to the airspace. "While the UAS is undoubtedly the way of the future, my concern must be on today, and right now, the era of the unmanned aircraft system in civilian airspace is just not here yet," Babbitt said. "When you're dealing with the FAA, remember that all of the components need to be addressed," he reminded the audience. "While air traffic will help you get from Point A to Point B, it's the aviation safety organization that sets the standards to be at Point A…To assist and be ready for UAS reaching maturity, we have special program offices in our aviation safety and air traffic organization. [W] e are doing what we can to help get you to market."[2] Under Huerta, those offices would be combined into the UAS Integration Office.

Babbitt's tenure as administrator would be abbreviated but eventful. Working alongside Transportation Secretary Ray LaHood, one of his first steps was to gather executives of airlines and pilot unions for a "call to action" to improve safety and pilot training in response to the crash of Colgan Air Flight 3407 near Buffalo in February 2009, three months before he arrived at the FAA. Controllers represented by the National Air Traffic Controllers Association were in the third year of a contract dispute, reporting for duty under FAA management imposed work rules and conditions. Assisted by mediators, the agency under Babbitt concluded negotiations with NATCA for a new contract in September 2009. Trouble ensued when multiple incidents of controllers sleeping on duty made national news, forcing the resignation of Henry P. "Hank" Krakowski, chief operating officer of the FAA's Air Traffic Organization, in April 2011. The ATO was the agency's largest component organization, responsible for nearly 80% of its operations budget.

Babbitt never knew funding stability as administrator. Democrats in the Senate and Republicans in the House clashed over a twenty-first short-term reauthorization bill in 2011, depriving the FAA of funding authority for nearly two weeks that summer. During the stalemate in Congress, the agency furloughed 4,000 employees and halted work on 258 tower construction and infrastructure projects worth $10.5 billion.[3] On a Saturday night that winter, police in suburban Fairfax City, Virginia, arrested Babbitt on suspicion of drunken driving; three days later he resigned, although a judge would later dismiss the charge. It was an unfortunate but honorable thing to do for the leader of an agency that claimed safety as its mantra. "Serving as FAA administrator has been an absolute honor and the highlight of my professional career. But I am unwilling to let anything cast a shadow on the outstanding work done 24 hours a day, seven days a week by my colleagues at the FAA," Babbitt stated in a press release announcing the resignation on December 6, 2011. "They run the finest and safest aviation system in the world and I am grateful that I had the opportunity to work alongside them."[4]

Huerta now owned the embattled FAA, although his promotion to administrator was not assured. In March 2012, Obama nominated him to a full five-year term to replace Babbitt, but South Carolina Republican Sen. Jim DeMint blocked the nomination, arguing that Republican presidential hopeful Mitt Romney should choose his own administrator if he defeated Obama in the November 2012 elections.[5] After Romney lost to Obama, DeMint withdrew his hold on the nomination and the Senate confirmed Huerta as FAA administrator on January 1, 2013.

Then fifty-six, Huerta was cut from a different cloth than Babbitt. The lanky, avuncular Babbitt—a Florida native—was a former Eastern Air Lines captain and Air Line Pilots Association president who hailed from the rank and file of airline pilots. Huerta, a Californian, was more reserved, shorter and squarer in stature, but with big hands and a firm handshake. He had the distinction of being the second Hispanic FAA administrator; the first was the storied aviator Elwood "Pete" Quesada, the man President Dwight Eisenhower appointed to lead the FAA when it was established in 1958. Huerta came to the job with a lengthy resume of management positions in the public and private sectors, including stints on the organizing committee for the 2002 Winter Olympics in Salt Lake City as an executive on loan from Affiliated Computer Services; as executive director of the Port of San Francisco; and as commissioner of the New York City

President Obama nominated Michael Huerta to serve as FAA deputy administrator, but Huerta was suddenly thrust into the administrator's role when his boss, Randy Babbitt, resigned in December 2011, after being arrested on suspicion of drunken driving. A judge later dismissed the charge. *Author photo*.

Department of Ports, International Trade and Commerce. During the mid-to-late 1990s, Huerta held executive positions in the Clinton administration under transportation secretaries Federico Peña and Rodney Slater, finishing as Slater's chief of staff. But he didn't have the dyed-in-the-wool aviation experience that Babbitt brought to the administrator's post. Sen. Maria Cantwell, a Washington Democrat, noted the deficit during Huerta's confirmation hearing before the Senate Commerce, Science, and Transportation Committee in June 2012. "When you came in front of this committee the first time, I recall there were many whispers about 'did you have enough background in aviation?' Well, you have demonstrated that that is a non-issue, and you have brought a great skill set, a great mindset to this job," Cantwell said.[6]

Indeed, Huerta was a practiced public servant who appeared unflappable during interrogations by various congressional panels and well-versed when explaining the technical nuances of NextGen, his original charge. The calm demeanor was useful at a difficult time. Three months into his term as the full-fledged FAA administrator, Huerta and other federal agency heads were compelled to make severe budget cuts under the arcane procedure known as "sequestration." Congress had inserted the budget-cutting mechanism in the Budget Control Act of 2011, but only as a doomsday option to force Democrats and Republicans to compromise on reducing the federal deficit. It convened a Joint Select Committee on Deficit Reduction comprised of potentates from both parties—popularly known as the "supercommittee"—that then spectacularly failed to reach agreement. As a result of the committee's failure, sequestration entered into force on March 1, 2013. For the FAA, that meant clipping $637 million from its budget in the seven months that remained in fiscal 2013. An Aerospace Industry Association study warned that sequestration could set back NextGen by a decade to 2035.[7]

The FAA considered unmanned aircraft to be a subset of NextGen; a new technology that it would weave into the future airspace system, much like commercial spacecraft, which was another industry on the rise. But UAVs were fast taking on a life of their own. While NextGen, with its four-dimensional trajectories and closely spaced parallel operations, was a difficult concept to communicate, "drones" were something the public could imagine. Surveyors, realtors, filmmakers, and journalists saw benefits in using them. The FAA, out of an abundance of caution or of bureaucracy, depending on one's viewpoint, seemed to be impeding progress toward introducing UAVs in US skies. Cornyn's hold on Huerta's appointment as deputy administrator was evidence of the mounting pressure on the agency at a high level to expedite approvals, and Congress had just legislated a new set of requirements and deadlines in the FAA Modernization and Reform Act. While he was still acting administrator, Huerta laid out the agency's case in a speech to the Unmanned Systems Conference in Las Vegas in August 2012. "As the provider of air traffic control, we must ensure the safety and efficiency of the entire airspace, including all people and property in the system—both manned and unmanned," he said, according to a text of the speech. "Our goal is to safely and efficiently integrate unmanned systems into our airspace."[8]

In the speech, Huerta acknowledged the FAA's internal structure needed to be realigned "to accommodate the great transformations that are happening. We need to work across agency departments to implement NextGen. And one of the most innovative aspects of NextGen is unmanned aircraft systems." Previously, different branches of the agency handled issues related to unmanned aircraft, but as of March 2012, the FAA had established an Unmanned Aircraft Systems Integration Office as "a centralized, one-stop portal" for everything related to civil and public UAS. "This office will develop a comprehensive plan to integrate unmanned aircraft systems and establish operational and certification requirements for UAS. It will also oversee and coordinate UAS research and development," Huerta said. "This office will think creatively because it will formulate new standards for new aircraft."

The idea for a single, consolidated UAS Integration Office—AFS-80 in the FAA's nomenclature—came from the executive ranks. It originated with John McGraw, then deputy director of the agency's Flight Standards Service, which fell within the FAA hierarchy under the Aviation Safety organization. The so-called integration office was the third office the agency formed to manage unmanned aircraft, McGraw remembered. Mindful of its experience in assigning people and resources to very light jets—another apparently hot new category of aviation that fizzled during the recession of the late 2000s—it moved cautiously. The first office resided within the Aircraft Certification Service, another part of the safety organization. When the government's use of unmanned aircraft by the Department of Defense, Department of Homeland Security, and NASA accelerated in the wake of the Iraq War, the FAA shifted the UAS office to the Flight Standards Service "in recognition that the demand was far outstripping our ability to put engineering standards in place to certify the machines," McGraw said. He would spend a lot of time on UAS issues with an executive on the FAA's air traffic side, Dennis Roberts, ATO director of airspace services. The two agreed that unifying Aviation Safety's Unmanned Aircraft Program Office and ATO's Unmanned Aircraft Systems

Group made sense. "Obviously the air traffic piece was huge, and we needed to have even more integration at the executive level, especially as Congress was formulating the reauthorization bill," recounted Jim Williams, the man the FAA would appoint to lead AFS-80.

Williams and McGraw had worked together in Atlanta and were friends outside of work. Williams was entering his seventh year as NextGen director of engineering services and was looking for a change. "When I heard that they were considering doing this, I'm like 'oh wow, perfect job for me,' and I let them know that I would be interested in doing it," he said. "That actually helped (McGraw) get the office stood up, knowing that there was somebody who could come in and take the job." Aviation Safety, the ATO, and ultimately Huerta signed off on the reorganization, positioning AFS-80 within the Flight Standards Service. The administrator, his chief of staff Sasha Johnson, and Michael Whitaker, who was sworn in as the new FAA deputy administrator in June 2013, backed its mission, Williams attested. "Absolutely, very much so," he said with conviction. Huerta "and Whitaker both are extremely supportive, and Sasha. They very much appreciate the work we're doing. It's helped them in that they've got one place to come to get the information they need, and we provide them a tremendous amount of service in dealing with congressional inquiries and testimony and public affairs—practically daily interactions."

The UAS Integration Office was not located at the FAA's Washington headquarters facility (the Orville Wright Federal Building) at 800 Independence Ave; rather, it was set back a block in L'Enfant Plaza, a complex Elwood Quesada's L'Enfant Plaza Corporation helped develop in the mid-1960s. The plaza complex was part of a matrix of ponderous, concrete government buildings—the Department of Energy and the Department of Housing and Urban Development among them—inspired by the Brutalist style of architecture popular at the time of their construction. Skirted to the north by CSX railroad tracks and to the south by the Southwest Freeway, this was the underbelly of federal Washington, a district that thousands of federal workers traversed by day but one that remained largely invisible to the throngs of tourists on the nearby National Mall.

On a Friday at mid-morning, Williams was dressed in a camel hair jacket and checked dress shirt, open collar. His third-floor office had a large window facing west, the view dominated by the Brutalist headquarters of the US Postal Service. In the corner was an azure-blue flag of the federal senior executive service; around his workspace was an assortment of photos and mementos and a scale model of the Insitu ScanEagle.

A stocky Tennessean with thick, graying hair and a prominent brow, Williams was true-blue FAA. He had been with the agency for all but the first four years of his career, when he worked for NASA and in private industry for Lockheed. While studying toward an aerospace engineering degree from Georgia Tech he enrolled in the university's cooperative education program, which placed him at NASA Ames Research Center for practical experience. He helped design the mount for the control stick pilots used in the Ames vertical motion simulator. Interested in the Space Shuttle program, he transferred to NASA's Johnson Space Center in Houston. There, as a "co-op" working in the mission control center, he participated in the Shuttle's first return to Earth in April 1981. After graduating from Georgia Tech in 1983, Williams went to work in the "back rooms" at mission control for two years.

An opportunity arose to work for Lockheed, which was flight testing the improved C-5B Galaxy airlifter for the US Air Force. Williams had strong feelings about leaving

Under pressure from industry and Congress to expedite entry of unmanned aircraft systems into the national airspace, the FAA consolidated its different program offices into the UAS Integration Office. Named to lead it was Jim Williams, a career FAA executive with experience in the agency's various branches. *Author photo.*

the nation's space agency. "I left NASA because I was pretty thoroughly disgusted with the Space Shuttle program management because of the numerous safety issues that were being ignored that all came out in the Challenger report," he said, referring to the 1986 post-mortem investigation of the Space Shuttle Challenger disaster. "I actually personally discovered a design flaw in the hydraulics that could have resulted in the Shuttle landing with the brakes on, which would have been catastrophic. Anyway, that's another story." He worked for Lockheed for two years in Marietta, Georgia, and under contract to Gulfstream Aerospace in Savannah for the flight test program of the GIV business jet. While working with Gulfstream on the civil program he came to know people from the FAA's certification branch. Earlier that decade, Lockheed had lost the competition to build the Air Force's C-17 airlifter and the proverbial writing was on the wall, Williams said. In 1987, he went to work in the FAA's Atlanta aircraft certification office.

Williams's FAA career included several years in Atlanta and Washington, DC, working for the certification branch. He left certification to work under associate administrator for research and acquisitions George Donohue on the "Capstone" program, which was evaluating the use of GPS-based aircraft tracking technology known as automatic dependent surveillance-broadcast. Under Capstone, the FAA supplied avionics to mail and charter planes operating in the challenging conditions of Alaska. Williams moved into communications procurement and then into air traffic control when the FAA folded the research and acquisitions organization into the newly formed "performance-based" ATO in 2003. In 2005, he became director of engineering services in the NextGen organization, the job he held when the UAS Integration Office came into being. "I've sort of done everything from detailed engineering oversight of airplanes all the way up through acquisition," Williams explained. "I had an understanding of the whole of the FAA and how it all works—the infrastructure, the air traffic, the certification. When this came along, I realized that there weren't many people in the FAA who actually had the broad background that could really understand the multitude of issues that unmanned aircraft were causing."

He started checking off the requirements Congress assigned the FAA in its 2012 reauthorization legislation. The FAA made an early deadline to expedite its process of issuing COAs to public safety agencies working with the National Institute of Justice, the research and development arm of the Department of Justice. "While the UAS Integration Office had already been working on a solution in advance of the bill, the 90-day deadline turned up the heat to get the solution completed," aviation attorney Timothy Adelman wrote in a column published by AUVSI. "Under its new leadership by James Williams...and in direct coordination with NIJ, the FAA has developed a streamlined COA process for public safety." Adelman added: "The news for public safety agencies and manufacturers is good. The FAA was once considered a major hurdle for public safety operations. While I still hear agencies and manufacturers claim that the barrier to small UAS operations is the FAA, that is no longer an accurate statement."[9]

Williams traveled to Alaska and interviewed oil companies on their plans for unmanned aircraft. His interactions with the FAA's UAS aviation rulemaking committee put him in contact with Paul McDuffee, Insitu's executive in charge of strategy and business development. Insitu had worked with the University of Alaska-Fairbanks on joint UAV flight demonstrations since 2006 and was known to ConocoPhillips. Williams forged an "other transaction agreement" with the oil company to operate Insitu's ScanEagle from one of its exploratory ships in exchange for sharing its operational data. On July 19, 2013, the FAA awarded restricted category type certificates to Insitu for the ScanEagle and to AeroVironment for the Puma, for the first time permitting operators to fly them in Alaska for commercial purposes and answering the congressional direction to open access to the Arctic. On August 5 the agency granted a COA to Insitu, which conducted the historic first commercial UAV flight from the ConocoPhillips-chartered *Westward Wind* on September 12 that year. On December 30, 2013, the FAA selected universities and other public entities in Alaska, Nevada, New York, North Dakota, Texas, and Virginia to operate the six UAS test ranges Congress required.

Separate from the provisions of the FAA reauthorization act, Williams participated in standards organization RTCA's Special Committee 203, which had been formulating UAS performance standards since October 2004. "I was overseeing the lack of progress of that group," he related, so he moved to restructure the committee of industry and government subject matter experts and accelerate its progress. In January 2014, Williams signed a memorandum of understanding with the Academy of Model Aeronautics to cooperate on the safe operation of model aircraft, which Congress had prevented the FAA from regulating. The agreement served to formalize the FAA's relationship with the hobby community. "We weren't focused in our efforts to integrate UAS. That's why I really supported creating this office—to help bring focus and attention," Williams recalled. "Instead of all these distributed efforts, we would have one focused, coordinated effort. Because it really is the biggest transformational change to aviation since the jet engine. I felt like the FAA really needed to get its act together to deal with it, because it was coming whether we wanted it to or not...That's how [the office] got stood up. It was obvious that it was needed."

There were also setbacks in the FAA's careful campaign to introduce UAVs. Just over a week before this author's visit to L'Enfant Plaza, an NTSB administrative law judge had dismissed a $10,000 fine the FAA levied against Raphael Pirker—the agency's first enforcement action that had advanced to a civil penalty—for piloting an electric flying wing over the University of Virginia campus for hire. The FAA alleged that Pirker flew the aircraft in a "careless or reckless manner," including through a tunnel, under an elevated walkway, toward people, and within 100 feet of a heliport. The judge found the FAA had no regulations that applied to a model aircraft, or that classified a model aircraft as an unmanned aircraft system.

Then, days after the judge handed down his decision in the Pirker case, a man brought his DJI Phantom 2 quadcopter to video the scene of a gas explosion in East Harlem that leveled two buildings and killed eight people. The man "captured stunning aerial images...thanks to his camera-equipped whirlybird," reported the New York *Daily News*, which featured some of the footage on its website. He told the newspaper he shot about thirty minutes of video before the Phantom's battery ran down and the police asked him to land the drone. "I mostly use it to shoot real estate or sports events. This was the first time I used it for breaking news," he was quoted as saying.[10] Had the radio frequency link to the quadcopter been lost or disrupted, he could have lost control of the aircraft at an emergency scene. "Ninety-nine percent of the people who are out there waving the flag, saying we want to use these things—they're not thinking about that, and they really need to," observed veteran FAA media handler Les Dorr, who was on hand for the Williams interview. "The industry is thinking about it, but in terms of the users, all of these guys that want to take their little Phantoms and fly them over Harlem, they're not thinking about that."

In a February 2014 report to Congress, the Department of Transportation inspector general found the FAA had completed eight of the reauthorization act's seventeen provisions relating to UAS, but was behind schedule in implementing the remaining nine. "Further, agency officials stated that FAA will not meet Congress' September 2015 milestone for safe integration of UAS but will complete some parts of integration—most likely for small UAS." Technological, regulatory, and managerial barriers were slowing progress. Although the FAA had established the UAS Integration Office in early 2012, "the office is not fully staffed and will have to reach out to FAA lines of business and offices beyond ATO, such as the Aircraft Certification and NextGen organizations. FAA has had difficulty working across lines of business in the past," the IG said. "Other organizational barriers limit FAA's oversight of current UAS operators," it added. "For example, regional UAS safety inspectors receive work assignments from the UAS Integration Office but report to their regional managers, resulting in competing priorities."[11] The Government Accountability Office also reported to Congress that month. The watchdog agency similarly noted progress, but warned of ongoing "uncertainty" over when the FAA would introduce UAVs into the airspace system. While the UAS Integration Office "did not have an operations budget, as of January 2014, the office has 33 full-time employees, and FAA is still finalizing agreements and other arrangements related to the reorganization, and it remains unclear what resources the office will have available to fulfill its role," it said.[12]

"I don't have a budget line item that says 'UAS' on it. I get my funding from multiple sources," Williams responded. One such source was the Flight Standards Service research budget—Williams's office had access to about half of the $8.4 million annual appropriation in 2014. That half had quadrupled from less than $1 million available for UAS as late as 2009. "It's been growing with the urgency of the situation; the budget has been moving in that direction," he said. Salaries for FAA employees and for paying contractors came from the agency's operations and facilities and equipment accounts. AFS-80 also had at its disposal some $2.5 million in staff years the FAA supported at MITRE Corporation, a federally funded research and development center.

On his office door, Williams had Scotch-taped copies of the historic type certificates the FAA's Los Angeles aircraft certification office issued in July 2013 for the Insitu ScanEagle and the AeroVironment Puma. He peeled off the copies and offered them to the author. "I've been meaning to take those down," he said. "It's kind of old news now."

CHAPTER 6

THE SMALL UAS RULE

In 2003, UAV manufacturer Aero Telemetry Corporation of Huntington Beach, California, was tasked to build several model airplanes for use in the flying sequences of *The Aviator*, director Martin Scorsese's rendition of the life of eccentric aviation pioneer Howard Hughes. Among those the company built were a half-scale model of the H-1 Racer monoplane that Hughes flew for a world speed record in 1935 and replicas of his twin-boom XF-11 reconnaissance airplane and the H-4 Hercules, better known as the *Spruce Goose*—a colossal seaplane with eight propellers that flew only once. Launched from a barge, the 275-pound, battery-powered version of the *Spruce Goose* flew several times over the water's surface in Long Beach Harbor, Hughes's original flight-test area. The XF-11 flew from the San Bernardino and Catalina Island airports. Released in 2004 and starring Leonardo DiCaprio as Hughes, *The Aviator* was nominated for eleven Academy Awards and won five. The film grossed $102 million at the box office in the United States and Canada and another $111 million overseas.[1]

Before the lines blurred between model aircraft and unmanned aircraft systems and before the rules firmed as to what they could and could not do, the FAA for more than a quarter century had managed model aircraft by providing guidance that operators were encouraged but not required to follow. The agency issued Advisory Circular 91-57, "Model Aircraft Operating Standards," on June 9, 1981. The one-page document stated: "Modelers, generally, are concerned about safety and do exercise good judgment when flying model aircraft. However, model aircraft can at times pose a hazard to full-scale aircraft in flight and to persons and property on the surface. Compliance with the following standards will help reduce the potential for that hazard and create a good neighbor environment with affected communities and airspace users." AC 91-57 recommended modelers choose a site that was sufficiently distant from populated and noise-sensitive areas, such as parks, schools, hospitals, and churches. They should not operate a model aircraft with spectators present until the model had been flight tested and proven airworthy. They should not fly a model aircraft higher than 400 feet above the ground and should notify air traffic control authorities when operating within three miles of an airport. They should cede the right-of-way and avoid flying near real aircraft, using spotters if possible.[2] "The advisory circular basically said 'use your head.' If you're flying a model airplane don't do it above 400 feet, and if you're within X miles of an aerodrome, notify the manager of that facility," said Gary Michel, who served as a regulatory attorney in the FAA's Office of Chief Counsel from 1979 to 2013. "It was an advisory circular; it wasn't a regulation. It was an attempt at a finesse to not go to a rulemaking—which, in hindsight, is really what should have been done—to try to educate the public that you shouldn't be using model aircraft for anything other than sport or recreation. We were doing the best we could at the time, given that there was so much inertia at the agency against rulemaking."

Michel started lobbying higher-ups at the FAA to craft a regulation on unmanned aircraft systems in the mid-1990s. During that period, Dutch national Peter Van Blyenburgh, who was trying to organize Europe's unmanned aircraft industry, visited the agency. He briefed Michel on his efforts, as well as two other FAA executives: Rick Hostetler, who served as air traffic services liaison to the Secret Service and the Department of Defense; and Stephen Isaacs, an air safety inspector with the office of Flight Standards. Around the same time, AUVSI wanted the FAA to issue advisory circulars on aspects of unmanned aircraft, such as their design, operation, maintenance, and pilot qualifications to help guide the growth of the US industry, which Michel advised against because there were no underlying regulations to link to the guidance. In 1997, Michel, Hostetler, and Isaacs put together a PowerPoint presentation on the FAA's need to regulate unmanned aircraft. They tried to convince as many senior level executives as they could, including Margaret "Peggy" Gilligan, who was then deputy associate administrator for Aviation Safety. Their argument was: "if the agency doesn't start to do something now, we're going to be behind," Michel said. "It really didn't go very far. Peggy said 'we have bigger fish to fry,' which was a fair statement because she had a lot more stuff on her plate in terms of the air carrier industry, general aviation accidents—runway incursions were a big thing. So they had their hands full. I still remember 'we have bigger fish to fry.'"

As the oil companies evaluated using small drones in the waters off Alaska in the mid-to-late 2000s, the level of interest and activity in the technology increased in the lower forty-eight states as well. Years later in 2014, Gilligan recalled that period during an airline industry conference. "I will be honest with you, it was the FAA's expectation that we would have manufacturers who wanted to bring this technology into the system

The FAA did not anticipate what a "game-changing technology" unmanned aircraft would become, Peggy Gilligan, the agency's associate administrator for aviation safety, acknowledged in 2014. Gilligan served as deputy associate administrator of the FAA's safety organization for fourteen years before being named to lead it in 2009. *Author photo*.

to apply for certification, because that was the standard model that we had," she said. "We expected that once we had an application, we would define the certification basis (and) they would build and manufacture to that certification basis. I think what we didn't anticipate sufficiently was what a game-changing technology unmanned systems really are. In fact, we have a manufacturing base that is not very familiar with the aviation approach to things, especially as you look at small UAS. That has added a challenge that we didn't anticipate." Nicholas Sabatini, who in 2001 was named Gilligan's boss as associate administrator of the FAA's Aviation Safety branch, created an unmanned aircraft program office in 2006 that was focused on safely introducing unmanned aircraft into the airspace system—a precursor of the UAS Integration Office. He appointed former air traffic controller Kenneth "Doug" Davis to lead it. According to Michel, the FAA started collaborating with Michael O'Shea, aviation technologies program manager with the National Institute of Justice, and Donald Shinnamon, who chaired the aviation committee of the International Association of Chiefs of Police—both early advocates of police use of unmanned aircraft. The agency was also aware that primarily real estate agents and developers were using small unmanned aircraft to photograph properties from the air and that some hobbyists were flying their models too close to airports or well above 400 feet. Regulators watched YouTube videos that hobbyists posted, showing their RC models flying in airspace used by airliners.

Before activity picked up, the FAA tacitly allowed unmanned aircraft operators to fly under the AC 91-57 guidelines for model aircraft; "that was the implication," AeroVironment's Ted Wierzbanowski remembered. When the rogue flights multiplied, the agency sought help from standards organization RTCA, which served the FAA as a federal advisory body to develop "best practices" that small unmanned aircraft operators could follow. An RTCA group of industry and government experts (Special Committee 203) had been at work since 2004, developing performance standards that would allow larger UAVs to fly in the same airspace with manned aircraft. Though smaller, closer-to-the-ground machines were not within its remit, SC-203 formed a subcommittee in 2006 to address them. Joseph Boyd, representing the Air Line Pilots Association, served as the subcommittee chairman. It held weekly teleconferences, during which subject matter experts were invited to share their ideas for best practices. "We started to hear some real scary stuff of really stupid things that people were doing," Wierzbanowski said. "The FAA guys and some of us that are professional aviators are on the phone going holy shit!"

On February 13, 2007, the FAA issued a "policy notice" clarifying its position on flying unmanned aircraft systems. The six-page notice began with a description: "Simply stated, an unmanned aircraft is a device that is used, or is intended to be used, for flight in the air with no onboard pilot. These devices may be as simple as a remotely controlled model aircraft used for recreational purposes or as complex as surveillance aircraft flying over hostile areas in warfare." There was a wide range of sizes and weights of UAVs, and they performed a variety of public services. "The one thing they have in common is that their numbers and uses are growing dramatically," the agency said. About fifty companies, universities, and government organizations were developing some 155 unmanned aircraft designs. There were three ways the FAA authorized operators to enter the airspace: public agencies required a COA; civil operators required a special airworthiness certificate in the experimental category; "and for model aircraft the authority is AC 91-57." Now it seemed that some operators were using the model aircraft guidance as the basis to fly unmanned aircraft commercially. "The FAA recognizes that people and companies other than modelers might be flying UAS with the mistaken understanding that they are legally operating under the authority of AC 91-57. AC 91-57 only applies to modelers, and thus specifically excludes its use by persons or companies for business purposes," the notice declared. It concluded by explaining the FAA had begun a safety review to "examine the feasibility of creating a different category of unmanned 'vehicles.' [T]he end product of this analysis may be a new flight authorization instrument similar to AC 91-57, but focused on operations which do not qualify as sport and recreation, but also may not require a certificate of airworthiness." Sabatini signed the document.[3]

Michel counted as one of his biggest successes convincing Sabatini that unmanned aircraft in their various manifestations were, indeed, aircraft within the agency's definition. In the 1980s, the FAA used the "legal fiction" that an ultralight vehicle was not an aircraft, reasoning that calling such a winged machine a "vehicle" would absolve it from aircraft registration and certification requirements, he said. By the late 2000s, small, remotely controlled aircraft, including even model aircraft, became a pressing challenge the agency needed to address. "I was saying as an attorney, look we have benignly neglected model aviation. Before the advent of modern technology—GPS, good electronics—if we had regulated kids playing with model aircraft on a Saturday morning on a playground, the media and Congress would have dope-slapped the agency." But the technology had evolved and was more affordable and accessible to people who were either unaware of or undeterred by the agency's governance of the airspace. "The FAA in my personal opinion had no choice but to shut everything down, and that's when they issued the clarification of existing policy which basically said you cannot fly small UASs for compensation and hire under the AC 91-57 rules," said Wierzbanowski. "So then, because of that, there was a big uproar."

Before the FAA banned commercial flights of unmanned aircraft, Patrick Egan was working as a remodeling contractor and supplementing the business by taking aerial photographs real estate agents wanted using a drone. "It was really the digital camera that revolutionized the concept," Egan said. "I was doing real estate shots. Then I got into doing condo high-rise projects. I would fly around and take pictures and basically they would do a mock-up of the view outside your window if you [were to] buy this unit. I would go out and make one thousand, two thousand bucks in a day and say 'this is easy money.'" He was flying a HobbyKing model airplane and planned to name the aerial photography business "Sky-Borg," somewhat tongue-in-cheek. "Before you charged money, it was a twenty-nine dollar RC aircraft called a Slow Stick," he said. "It's really, really lightweight. I did a lot of urban flying. If you did have a mishap there would be no issues." Egan was a member of the Remote Control Aerial Photography Association and participated on the RTCA subcommittee on small unmanned aircraft. Months before it happened, FAA officials forewarned him that the agency planned to clamp down on commercial flights, but they offered assurances that the FAA would propose another regulation within sixty to ninety days to accommodate them. "I stopped doing it because I was interfacing with the FAA," Egan said. "A lot of people backed off, and mainly those were people that had pilot's licenses. So people stood down. In my estimation [the FAA] created a very unsafe situation. The people that knew the rules—they actually cut them right out."

In April 2008, acting FAA Administrator Robert Sturgell signed an order that formally established a small UAS aviation rulemaking committee—an ARC, for short—

to make recommendations the agency might use to develop regulations governing small unmanned aircraft. The order stated:

> The law enforcement and aerial photography industries, plus others conducting remote sensing activities, have mistakenly interpreted FAA advisory circular 91-57, Model Aircraft Operating Standards, for permission to operate small UAS for research or compensation or hire purposes. To promote safety while updating and clarifying guidance, we at the FAA will conduct a formal safety analysis of small UAS, and then promulgate federal regulations for their design, operation and registration.[4]

Doug Davis, manager of the unmanned aircraft program office, invited Wierzbanowski to serve as the ARC's industry co-chairman. Bruce Tarbert, who reported up through Davis as the FAA's national airspace system integration team lead, was the government co-chairman. Among the members were Aerosonde inventor and Insitu co-founder Tad McGeer, now representing his new company, Aerovel Corporation; Paul McDuffee of Boeing Insitu; Michael O'Shea of the National Institute of Justice; Donald Shinnamon of the International Association of Chiefs of Police; officials with the Department of Defense and NASA; and representatives of the Aircraft Owners and Pilots Association, the Air Line Pilots Association, the Academy of Model Aeronautics, AUVSI, and standards organization ASTM International. After "hitting the congressmen hard" to participate on the ARC, including his own representative, California Democrat Doris Matsui, the outspoken Egan also gained a seat at the table, representing the Remote Control Aerial Photography Association.

The twenty-member committee delivered recommendations in April 2009. It organized UAVs into four groups ranging in weight from 4.4 pounds or less to 55 pounds—the outer limit the Academy of Model Aeronautics set for small model aircraft. It reserved a fifth category for lighter-than-air systems but did not make specific recommendations. The ARC prescribed operational limits and required capabilities for each group; for example, the smallest air vehicles were limited to flying 400 feet above ground level, more than three miles from an airport, and within the operator's visual line of sight. They would be "frangible" (breakable)—constructed of paper, wood, or plastic, with no substantial metal parts to minimize the risk of injury or damage in the event of a crash. The pilot in command of the smallest group would not need a pilot's certification, but would have to complete an approved training course. No group would be allowed to fly in Class B airspace surrounding the nation's busiest airports.[5] Not everyone on the committee was pleased with the results. Though expressing admiration for Wierzbanowski, Egan believed the Department of Defense vendors and other powerful interests represented on the ARC designed the rules in their favor. "We were the only people that came in with anything. We came in with community-based guidelines, operating procedures, insurance, a testing program. We gave it to (the FAA) and they didn't even read it," he said. "I groused for 10 months that we were getting robbed, how special interests got what they wanted." The 400,000-member Aircraft Owners and Pilots Association (AOPA) and the Air Line Pilots Association "rode roughshod" over the deliberations, he charged. "If they didn't like something, it didn't happen. AOPA has got (FAA) administrator-on-the-phone after hours pull, and that's just the facts." When the committee's term expired, ARC members expected the FAA would formally comment on the recommendations and begin crafting a notice of proposed rulemaking—a draft rule called an NPRM—almost immediately. "The expectation was that roughly a year from receiving the recommendations from the ARC, the NPRM would in fact be published," said McDuffee.

Promulgating a federal regulation was a layered process. Once the FAA deemed a regulation was necessary, it developed the proposed rule in-house and sent it forward to the Department of Transportation for approval. The DOT vetted the rule and sent it to the White House Office of Management and Budget. The OMB's Office of Information and Regulatory Affairs reviewed it to ensure the rule was "consistent with applicable law [and] the President's priorities" and did not conflict with other agencies' policies.[6] The FAA's next step after the rule cleared OMB was to publish it in the *Federal Register* as an NPRM for public comment. After considering public feedback and making any amendments, the agency would publish a final rule about eighteen months later. The FAA initiated the rulemaking effort for the small UAS regulation on July 28, 2009, according to the DOT's regularly updated report on significant rulemakings. The draft rule was originally scheduled for publication on March 10, 2011. In late 2014—more than seven years after the FAA banned using unmanned aircraft for business purposes until further notice, and more than four years after the small UAS aviation rulemaking committee delivered its recommendations—the agency had still not published an NPRM. The DOT, in boilerplate fashion, cited "unanticipated issues requiring further analysis" for causing the delay.

A common refrain at industry gatherings came to be: When will the FAA publish the small UAS rule? The *sUAS News* website, which Egan edited, featured the "FAA Miss-O-Matic" ticker to count the days since the rule was due. There were real consequences of the delay: the California Film Commission—a state agency that issued permits to production companies to film on state-owned property—announced in May 2011 that it would no longer grant permits to companies that used unmanned aircraft for either film or still photography. The use of remote-controlled aircraft fitted with cameras for commercial purposes "is not yet regulated and therefore not authorized by the FAA," the commission stated in a production alert. "We as a state agency couldn't issue a film permit when people wanted to use these because the FAA said they couldn't do so. Lots and lots of production companies were interested in using these devices," said Amy Lemisch, the commission's executive director.

Draft Regulation "Warehoused"

There were "at least four false starts on rulemaking" within the FAA dating to around 2003, according to Michel. The agency assigned "well-intentioned" people to write drafts of the small UAS rule that kept getting bounced back to them as unacceptable, never making it beyond the DOT review stage. Successive "barrels" were thrown in the rule's way, from inside and outside the FAA. At one point, a team working on the rule was sidetracked when the agency chose its COA process to be an ISO 9000 quality management system compliance project. "They picked unmanned aircraft to be the poster project. We dropped any forward-looking rulemaking effort and sat around for months developing an ISO flow chart of how you get a COA. It's that kind of thing. There was always something that indicated to me they don't really care about this project." In his long career with the agency, Michel said he was "always amazed at how many rulemakings—forget unmanned aircraft, all the other rules—how many times you'd get to rulemaking status meetings and the senior executives had not read the document." In his opinion, the FAA lacked the leadership to send the rule back to the DOT "and say we're done." It was the textbook example of a "captured agency," one that accommodated major players like Boeing and the airlines in setting regulations but dragged its feet when confronted with an upstart industry—small unmanned aircraft—that actually wanted to be regulated to gain legitimacy.

Those were among internal barrels Michel saw slowing the release of the small UAS rule. From the outside, the FAA encountered an unexpected firestorm over the privacy threat of drones fitted with sophisticated cameras and possibly even firearms or bombs. "We were getting pinged a lot by Congress on privacy and on concerns about UAS weaponization," Michel remembered. "Public interest groups that are privacy based raised the specter that we (as a society) are over-surveilled, that UAS are Orwellian, the totalitarian state—all of that stuff. I get the sense that both the White House Office of Management and Budget and DOT warehoused the rule in the FAA. I think they figured if it's got to be on someone's desk with all of these privacy concerns swirling around, why should it be DOT, why should it be OMB?"

John McGraw, with the agency's Flight Standards Service, recalled first seeing a draft of the regulation in the mid-2000s. He described its provenance somewhat differently. As he later told one conference audience, the small UAS regulation presented the FAA with a "big, hairy problem" because of its broad scope and the different types of operations and aircraft it sought to manage. "It was not workable," McGraw said of the first cuts of the rule. "It was overly complex, overly onerous for both FAA and industry. It was not going to work out of the gate, so the team was sent back to start over again essentially—twice at least. Part of that is [the industry] was new and nobody was quite sure how it would be managed or how big it would be... When you have a rapidly changing technology it's hard to do rulemaking. It's [like doing] surgery on a beating heart."

Even with the draft regulation mired in the federal bureaucracy, the "airspace integration" effort advanced on other fronts. After disbanding the small UAS ARC, the

In 2012, the AUVSI Foundation gave its outstanding contributor award to Ted Wierzbanowski (left), a former Air Force colonel and experimental test pilot who worked for AeroVironment. At right is Daryl Davidson, the foundation's executive director. Wierzbanowski wore many hats in the effort to integrate unmanned aircraft. *Courtesy of Robert Bruni, Ambience Photography*

FAA inquired if standards organizations would be interested in developing guidelines for the design, manufacture, maintenance, and other aspects of small unmanned aircraft systems. It selected for the task ASTM (the American Society for Testing and Materials). "That's when we really started to get energized to develop the standards for small UASs," said Wierzbanowski, who was wearing yet another hat as chairman of ASTM's Committee F38 on Unmanned Aircraft Systems. "Once the FAA sort of sanctioned ASTM to do this, then ASTM got a lot more membership and a lot more interested folks working on it." The organization announced in April 2014 it had approved seven new standards.[7]

On another front, FAA Administrator Randy Babbitt signed an order in June 2011, creating the Unmanned Aircraft Systems ARC, focused on "operational approval standards, operating procedures, complete aircraft system capability and guidance material required for UAS operations." The associate administrator for Aviation Safety—by then, Peggy Gilligan—had "sole discretion" to appoint members to the committee who represented the aviation community and to decide when its recommendations were released to the public. The agency did not readily reveal the new ARC's membership.

One person who was not invited to participate was Egan, who filed a Freedom of Information Act request seeking the names of the people who were asked to serve. The FAA responded in October 2011 with a list that included some participants from the earlier ARC, including McDuffee, Wierzbanowski, O'Shea, AOPA, and the Air Line Pilots Association, as well as new companies and associations.[8] Egan's remote-control aerial photographers association was eventually invited to participate on a "Beyond Visual Line of Sight" committee the FAA created to advise the ARC. The ARC's first task was to advise the FAA in drafting its "Integration of Civil Unmanned Aircraft Systems in the National Airspace System Roadmap," which the agency unveiled in November 2013. "We also put together an implementation plan on exactly how to make that tick" containing some 640 tasks, said McDuffee. "Unfortunately the implementation plan was never released. The FAA still has it; they have chosen not to act on it." As of the summer of 2014, McDuffee remained disappointed that the agency had not embraced the plan. "This is no small project," he advised. "I've heard it likened to the same level of activity needed to put a man on the moon. There are that many intricacies and little twists and turns in the path that all need to be dealt with before we're going to be successful to the point where we have routine access to the airspace. It's a big deal." In June 2014, FAA Administrator Michael Huerta signed an order that extended the ARC's charter another two years to 2016. Attached to the order was a list of ARC member organizations that included some of the industry's biggest names: AeroVironment, Boeing Insitu, General Atomics, Lockheed Martin, Northrop Grumman, Raytheon, Rockwell Collins, and Honeywell.[9]

The Pirker Case

For much of 2014, the FAA's authority over the nation's airspace from the ground up was in doubt, a situation exacerbated by the absence of a small UAS regulation. The agency brought the trouble upon itself. A year earlier, it had levied a $10,000 fine against Raphael Pirker, a twenty-something Swiss national, for allegedly recklessly flying a Ritewing Zephyr over the University of Virginia campus in Charlottesville, Virginia, on October 17, 2011. While the FAA had issued warnings to other operators of small drones before, this was its first such "enforcement action" to rise to the level of a civil penalty. In an assessment order, it said the foam construction Zephyr flying wing was an unmanned aircraft system and that Pirker flew it "in a careless or reckless manner" in violation of federal aviation regulations Section 91.13(a), a rule that applied to the operators of all small non-commercial aircraft. Further, the agency said Pirker conducted the flight for compensation by taking aerial photos and video of the UVA campus and medical center for a marketing agency. Though the order did not state as much, operating an unmanned aircraft for hire contravened the agency's February 2007 policy notice banning flying drones for business purposes.

Known by the nickname "Trappy," Pirker was considered a pioneer of the "first-person view" method of aerial videography. When flying FPV, the operator of a remote-controlled aircraft—either wearing goggles or using a video monitor—sees the same view as the camera on board the aircraft. Pirker's Team Black Sheep had filmed a dramatic video of New York City in 2010 that revealed a bird's-eye view of bridges, buildings, and the Statue of Liberty. As of 2014, the video had generated more than two million views on YouTube. The crew also produced aerial footage of the Golden Gate Bridge in San Francisco, the Grand Canyon, the Hoover Dam, and the music-synchronized water fountains of the Bellagio resort in Las Vegas, as well as other US and international landmarks. But it was the flight over New York that helped land Pirker an assignment from Lewis Communications to film the UVA campus and, he believed, attracted the FAA's attention. "It wasn't really the flight at Virginia that got me in trouble, it was the New York flight. That was purely recreational," Pirker told the author. "We just went on vacation in New York. We tried to be tourists and visit all the cool spots, but by drone. Back then it wasn't a normal thing to do." The FAA originally notified Pirker in

Raphael Pirker is shown with a Ritewing Zephyr flying wing. A Swiss national, Pirker was considered a pioneer of the "first person view" method of aerial photography. He and his Team Black Sheep crew filmed dramatic video of landmarks in the United States and other countries. *Anonymous*

April 2012 that it intended to assess a fine for the UVA mission, then more than a year later FedExed it to him in Zurich. In an on-line forum hobbyists used, Pirker asked if any lawyers could advise him about US aviation law. The appeal brought him to the attention of New York City attorney Brendan Schulman, who was himself a model aircraft enthusiast. "I had actually been following the FAA commercial prohibition for a while just out of interest as a lawyer and as someone who had a passion for the technology," said Schulman, who started flying a Hobbico Superstar 40 as a teenager and figured he owned thirty model aircraft. "I knew for a number of years that there was a purported ban on commercial use of model aircraft, but never any regulation or citation to what legal authority stood behind that prohibition. I knew it contradicted all sorts of model aircraft use in commercial context."

Schulman filed a motion to dismiss the FAA's action against Pirker with the National Transportation Safety Board, which served as a court of appeal for airmen, mechanics, and mariners who were penalized by the FAA or the Coast Guard. The FAA's action was "an improper attempt…to impose an unprecedented civil penalty upon the operator of a five-pound radio-controlled model airplane constructed of styrofoam," the motion stated. The agency sought to fine Pirker under federal aviation regulations Part 91 General Operating and Flight Rules, "notwithstanding the fact that it has never promulgated any enforceable regulation concerning the operation of model airplanes, and on the contrary has for decades expressly avoided such regulation and enforcement." Its only guidance that related to model aircraft was AC 91-57 (of 1981 vintage), which provided voluntary operating guidelines and made no distinction between model aircraft flown for commercial purposes and those flown for recreation.

In 2007, without undertaking the requisite notice-and-comment process, the FAA articulated new regulations couched as a policy notice. It began by reclassifying model

Attorney Brendan Schulman won the first round when he challenged the FAA's penalty against Raphael Pirker, casting doubt on the agency's regulatory authority. A National Transportation Safety Board administrative law judge granted Schulman's motion to dismiss the $10,000 fine, but the NTSB board reversed the judge's decision. *Author photo.*

aircraft as unmanned aircraft systems, then announced that such aircraft could not be used for business purposes. Model aircraft that were flown commercially required the FAA's specific authorization and were subject to federal aviation regulations. "The 'business purpose' distinction has no basis in any regulation or in the voluntary AC 91-57 standards. It is a new legal prohibition issued by the FAA," the motion argued. "The sole question presented here is whether there *currently* exists any valid regulation that would authorize the FAA to impose a monetary fine upon a model aircraft operator. The answer is plainly 'no.'"[10]

Some five months after Schulman filed the motion, NTSB Administrative Law Judge Patrick Geraghty found in Pirker's favor and dismissed the FAA's complaint against him. In his decisional order dated March 6, 2014, Geraghty wrote that disputing the facts of the FAA's reckless operation charge against Pirker "must be deferred, pending resolution of the threshold issue of (the FAA's) authority to exercise FAR regulatory action over model aircraft operations." Memorably, he argued that accepting the FAA's interpretation of what constituted an aircraft subject to regulation "would lead to a conclusion that those definitions include as an aircraft all types of devices/ contrivances intended, for, or used for, flight in the air. The extension of that conclusion would then result in the risible argument that a flight in the air of, e.g., a paper aircraft, or a toy balsa wood glider, could subject the 'operator' to the regulatory provisions of FAA Part 91, Section 91.13(a)." When Pirker flew his Zephyr over the UVA campus there was no "enforceable" FAA regulation applicable to model aircraft or for classifying a model aircraft as a UAS," Geraghty found.[11]

The FAA appealed the decision to the full NTSB board within days, but not before the media proclaimed the agency unequipped to regulate model airplanes or commercial drones. "Here's Why Commercial Drones Just Got Legalized," Jason Koebler argued in the on-line magazine *Motherboard*. Koebler allowed that "a final decision isn't likely to come that easy," but for the moment "drones are essentially legal based on a technicality…because the FAA was complacent or incompetent, depending on who you ask."[12] The website *News Ledge* reported: "An administrative judge with the NTSB has ruled that the FAA lacks the jurisdictional authority to ban the use of commercial drones…For now the ban has been lifted. Want to resume your planned drone delivery service? Go for it."[13] FlowerDeliveryExpress.com, an online florist, announced via PRNewswire that as a result of Geraghty's ruling it would resume testing drones to deliver flowers in the Detroit metro area. The *Wall Street Journal's* headline, "Judge Says FAA Lacks Clear Authority on Commercial Drones," was careful in assessing the import of the ruling, even if the newspaper's on-line coverage featured an Associated Press photo of a Customs and Border Protection Predator B.

At AUVSI's Unmanned Systems conference that May, aviation safety inspector Dave Morton told a roomful of policemen the FAA had no intention to stand down as a result of the Pirker decision. "Unfortunately, this particular activity has created in the blogosphere and in the media and in lots of other places this idea that the FAA is impotent to do anything about unauthorized operations, which is not true," he said. "We are taking enforcement action against those (incidents) that we think are egregious." Battle lines formed for the NTSB's review of Geraghty's ruling. Six parties filed "friend of the court" amicus briefs: four supporting Pirker's position that the ruling be upheld and two backing the FAA's contention that it should be overturned. Notable among the pro-Pirker group was the "News Media Amici," a coalition of major news organizations including the Associated Press, *The Washington Post*, *The New York Times*, Cox Media Group, Hearst Corporation, and Gannett. They contested the FAA's regulation of drones on freedom of speech grounds. The National Agricultural Aviation Association, an organization that represented "aerial applicators"—pilots and businesses that used aircraft to spray pesticides and seed on farmers' fields—was in the FAA's corner. The other pro-FAA party consisted of two former agency officials: John McGraw, who had been deputy director of the FAA Flight Standards Service; and Nicholas Sabatini, who signed the seminal 2007 policy notice that banned using unmanned aircraft for business purposes.

With the Pirker case still pending before the NTSB board, Brendan Schulman arrived for a meeting at the offices of Kramer Levin Naftalis & Frankel in midtown Manhattan carrying a Spectra flying wing under one arm and a miniature Blade Nano QX quadcopter in his free hand. Schulman was rail thin, erudite, with thinning dark hair and glasses. The Blade Nano was for kicks—working a hand controller, he performed a skilled flyover of the conference room table, briefly entrusting the little buzzing machine to the author's unsteady hand. The Spectra related to a second legal challenge Schulman mounted against the FAA's attempts to regulate small unmanned aircraft a month after Geraghty's ruling in the Pirker case. The agency ordered Texas EquuSearch Mounted Search and Recovery Team—a nonprofit search-and-rescue organization that had coordinated volunteer searches in hundreds of missing persons cases since its founding—to stop using radio-controlled airplanes fitted with cameras in its efforts. Schulman filed a petition for review of the order with the US Court of Appeals for the District of Columbia Circuit. Also named in the petition was Eugene Robinson, a co-founder of the Remote Control Aerial Photography (later Platform) Association who ran a nonprofit aerial search operation and volunteered his services to Texas EquuSearch. "Whatever you might think of Raphael's style of flying there was nothing objectionable about what Texas EquuSearch was doing," Schulman said. "They always coordinate (flights) with the incident commander, they aren't flying near airports, it's all line of sight, it's not FPV."

In July 2014, the appeals court simply dismissed Schulman's petition, explaining that the warning the FAA sent to Texas EquuSearch, which an aviation safety inspector communicated by email, did not represent a formal cease-and-desist order and had no legal consequences requiring the court's review. Both the FAA and Schulman claimed advantage from the decision. Texas EquuSearch was later able to deploy a catapult-launched MLB Super Bat drone to help search for a missing woman in Plano, Texas. The permission came by a roundabout route: the FAA granted an emergency COA to the National Institute of Standards and Technology—which used the aircraft for fire research—at the request of the Plano police. Robinson, who also worked for NIST on the fire project, piloted the aircraft.

Meanwhile, Schulman filed petitions in the DC circuit court seeking its review of an FAA interpretation of the special rule for model aircraft that Congress established as Section 336 of the FAA Modernization and Reform Act of 2012. The petitions, later consolidated, represented the interests of the Academy of Model Aeronautics, an organization of research universities named the Council on Governmental Relations, a group consisting of the UAS America Fund, which claimed access to $2.2 billion in private financing, and others. Schulman summarized their concerns over the FAA's interpretation: it was overly narrow in defining what it meant to be a hobbyist; it called into question hobbyists' use of FPV goggles by precluding their use for line-of-sight operations; and it imposed new restrictions that didn't exist before, such as requiring hobbyists to obtain permission before flying within five miles of an airport, as opposed to just notifying the airport.

But *Huerta v. Pirker* remained his highest-profile case. Schulman did not subscribe to Pirker's belief the FAA was retaliating against him for the New York City flight by penalizing him in Virginia. "I think they have had Raphael on their radar screen for awhile because he's a pioneer and he was flying in places that they might consider objectionable. My understanding with the New York flight, if there was an investigation, they concluded they couldn't do anything. In Virginia, the one difference is that the flight had a commercial purpose," Schulman said. The FAA faced a drone outbreak and Pirker and his Team Black Sheep probably stood out in the crowd. "I think the agency was interested in making an example out of somebody," he theorized. "They knew about (Pirker) from his other activities. They also knew that he was a foreigner and probably would be unable to defend himself because he wasn't based in the United States. I think they probably thought it would be an easy case to set an example and it would stand for the proposition that a commercial operation could result in a fine."

In a decision it announced on November 18, 2014, the NTSB, under acting chairman Christopher Hart, reversed Geraghty's ruling, siding with the FAA that Pirker's Ritewing Zephyr was indeed an aircraft subject to regulation against unsafe operation. The definitions of aircraft contained in federal law "are clear on their face," the board said. "Even if we were to accept the law judge's characterization of (Pirker's) aircraft...as a 'model aircraft,' the definitions on their face do not exclude even a 'model aircraft' from the meaning of 'aircraft.' Furthermore, the definitions draw no distinction between whether a device is manned or unmanned. An aircraft is 'any' 'device' that is 'used for flight.' We acknowledge the definitions are as broad as they are clear, but they are clear nonetheless." Notably, the board did not address the FAA's prohibition against using unmanned aircraft for business purposes dating to the 2007 policy notice. It remanded the case to Geraghty, instructing the judge to conduct a "full factual hearing" on whether Pirker flew the Zephyr recklessly.[14] Under Schulman's guidance, Pirker and the FAA would eventually settle the charges in January 2015. Pirker agreed to pay $1,100—a fraction of the FAA's original civil penalty—without admitting to wrongdoing.

Still under "executive review" as 2014 wound to a close was the FAA's promised notice of proposed rulemaking for small UAS. The latest version of the DOT's significant rulemaking report at the time indicated the FAA would release the draft regulation on December 22. But instead of unveiling the long-awaited rule that day, the FAA joined with trade organizations—AUVSI, the Academy of Model Aeronautics, and the newly formed Small UAV Coalition—to announce a "UAS Safety Campaign." With the holiday weekend approaching, they expected thousands of new hobby drones gift-wrapped and placed under Christmas trees would soon join the nation's gathering fleet. The campaign featured a website and a YouTube video explaining the FAA's guidelines for recreationally flying drones. The campaign motto was "Know Before You Fly." It was a noteworthy demonstration of industry-government cooperation that was nevertheless dampened by an article *The Washington Post* published on-line the prior evening and on page 1 of its print edition the day of the announcement.[15]

Based on apparently leaked "documents and emails," the lengthy article revealed senior FAA executives had overruled aviation safety inspectors who opposed the Section

333 exemptions the agency granted to Hollywood-affiliated aerial filmmakers. A drone operated by one of those companies—Pictorvision—had already flown off a movie set in California and disappeared. With the number of applications for exemptions mounting, the agency was leaning on employees to expedite approvals. "Several FAA employees, speaking on the condition of anonymity for fear of retaliation or losing their jobs, said supervisors are pressuring analysts to rubber-stamp the applications without a rigorous safety review," the *Post* charged. In an apparent conflict of interest, for a time the FAA had indirectly employed John McGraw, its former deputy director of flight standards, to help review applications, even though McGraw consulted for the Hollywood filmmakers and other companies seeking exemptions. Pressed by the newspaper to respond, the FAA said that it stopped using the contractor—SeaTec, an engineering consulting firm—for which McGraw worked. "I was working for a company that decided to help the FAA with processing the volume of applications. We had safeguards in place so I couldn't access anything," McGraw told the author. "It was mostly a misunderstanding." After the *Post* article ran, "they changed their mind and said no, we don't want you."

On another front, Kramer Levin announced that Brendan Schulman had filed a formal petition for rulemaking with the FAA on behalf of the UAS America Fund. The petition called on the agency to forgo the lengthy NPRM process and implement a "final direct rule" governing the commercial operation of "micro" unmanned aircraft weighing three pounds or less. It proposed that such machines be flown below 400 feet, at least five miles from an airport, and by "pilots with a demonstrated level of aeronautical knowledge." Supporting the petition was a study the fund and Kramer Levin commissioned. Analyzing twenty-five years of FAA bird strike data, the study concluded that similarly sized drones flown as the petition proposed would not pose "a significant increase in risk to manned aircraft."[16]

The FAA finally announced the small UAS draft regulation nearly four years later than expected on February 15, 2015, a Sunday during the President's Day holiday weekend. The agency was apparently rushed into making the announcement after an economic analysis of the regulation was inadvertently published on the government website Regulations.gov. The document was quickly removed, but not before it made the rounds on Twitter and *Forbes.com* and other media reported its findings. The FAA scheduled a teleconference at ten o'clock in the morning to unveil the rule, but had to change the call-in number when the conference line malfunctioned. When Transportation Secretary Anthony Foxx and Administrator Michael Huerta opened the teleconference forty minutes later than originally scheduled, they delivered some surprising news the leaked document hinted at. The FAA's proposed regulation would be less restrictive and burdensome than commercial drone aspirants feared. Under the agency's new "Part 107" rule for non-hobby, non-recreational uses of drones weighing less than fifty-five pounds, operators would be required to fly the machines within their visual line of sight—unaided by any device other than corrective lenses—to a maximum altitude of 500 feet above ground level, a maximum airspeed of 100 miles per hour (87 knots), and only during daylight hours. Flights in airspace sectors other than Class G, which the agency designated as low-lying airspace that was not otherwise classified or controlled, would require permission from local air traffic control authority. The draft regulation negated beyond line-of-sight flights, flights over "persons not directly involved in the operation," and "external load" operations—limitations that were problematic for newsgathering, package delivery, aerial farming, and other applications of drones. The

FAA took the advice its small UAS aviation rulemaking committee had delivered in 2009, inserting in the draft regulation a “micro UAS” classification that would permit more flexible operation in Class G airspace for drones weighing 4.4 pounds or less.

The proposal was remarkable for what it did not require. Small drone operators would have to be at least seventeen years old and vetted by the Transportation Security Administration—the same federal agency that staffed security checkpoints at airports—to determine whether they posed a security threat. But they would not need a private pilot’s license or a medical certificate to fly their machines, things the FAA required of Section 333 exemption holders. “Rather than requiring a private pilot license, we propose that operators obtain a newly created FAA unmanned aircraft operator certificate by passing a knowledge test focusing on the rules of the air,” Huerta told the teleconference. The certificate would come with a small UAS rating, and operators would have to renew it every two years by passing a written proficiency test. “You can think of it as sort of an endorsement to your driver’s license,” the administrator added. “The basic thing that we’re trying to establish here is this is fundamentally different than being a private pilot. A number of requirements that might pertain to being a private pilot simply don’t apply when you are flying an unmanned aircraft. What does apply is your ability to operate within airspace with other aircraft.” And the machines that operators used would not require an FAA airworthiness certificate, a potential show stopper for small drone manufacturers. “Such a certificate could take a manufacturer between three to five years to obtain. With the pace of innovation in the market, an unmanned aircraft could very well be outdated by the time it obtained a certificate,” Huerta said.

Aviation industry groups and other interested parties were generally positive, though cautious in their reactions, saying they needed more time to study the NPRM. In a string of tweets Patrick Egan—no FAA cheerleader—proclaimed that “many selfless people” had put years of work into the effort to introduce drones into the national airspace system. “Even the folks at the FAA deserve a THANK YOU!” he conceded. When the public comment period for the proposed regulation closed in late April, the FAA received 4,597 responses—a substantial number, but not the tens of thousands of comments some had anticipated.

Overshadowed by the FAA’s release of the small UAS draft regulation, the same day the White House issued a presidential memorandum signed by President Obama. Under the heading, “Promoting Economic Competitiveness While Safeguarding Privacy, Civil Rights and Civil Liberties in Domestic Use of Unmanned Aircraft Systems,” it required federal agencies provide public notice of their use of drones and make available on an annual basis a summary of their UAS operations during the previous fiscal year. As with information the government collected using any technology, the information it gathered on individuals using drones as a platform was subject to the same regulations and policies for collection, usage, storage, and dissemination. The memorandum called on the National Telecommunications and Information Administration, a branch of the Department of Commerce, to lead a “multi-stakeholder engagement process” that included private-sector participation. The effort would “develop a framework regarding privacy, accountability and transparency for commercial and private UAS use,” the document stated. Indeed, concern over the surveillance capabilities of drones was another looming challenge operators faced.

CHAPTER 7

PRIVACY AND FREE SPEECH

On January 15, 2014, the Senate Committee on Commerce, Science, and Transportation, chaired by West Virginia Democrat John D. "Jay" Rockefeller IV, held a hearing on Capitol Hill entitled "The Future of Unmanned Aviation in the US: Safety and Privacy Considerations." The first witness to testify before the senators, who were arrayed in a half circle at the front of the ornate hearing room, was one of their own: Sen. Dianne Feinstein of California. The esteemed eighty-year-old one-time mayor of San Francisco came as a special witness. After exchanging niceties with Rockefeller she delivered a lengthy opening statement. "I believe civilian drone technology, much of which has been developed in California, has great potential for both beneficial uses and for job creation. But the unique capabilities of the drone bring with it significant risks, most notably related to privacy and public safety," said Feinstein. She argued that Congress should proceed with caution before allowing the commercial use of unmanned aircraft, and set "reasonable rules" to both protect the American people and foster an emerging industry. As chairman of the Senate Intelligence Committee, she had seen firsthand the surveillance prowess of drones—not that civilian drones would be the same as those used for overseas national security operations, of course.

Breaking away from her written statement, Feinstein then told a personal story. There was "a demonstration in front of my house and so I went to the window to peek out and see who was there, and there was a drone right there at the window looking out at me," she said, holding the palm of her right hand near the tip of her nose. "Obviously, the pilot of the drone had some surprise, because the drone wheeled around and crashed, so I felt a little good about that." Feinstein, who was privy to some of the nation's most closely guarded secrets, had been buzzed at home by a drone—apparently a toy helicopter flown by activists with Code Pink, the sworn enemy of drone strikes.[1] Indeed, the antiwar organization was behind the June 15, 2013, incident at Feinstein's San Francisco mansion, Medea Benjamin confirmed to the author sometime later. She pulled up an Internet photo of the offending aircraft—a V-MAX X-Power radio-controlled mini helicopter that retailed for around thirty dollars.

Feinstein's intelligence committee was responsible for overseeing the CIA, and as its chairman she was controversial for seemingly low-balling civilian casualties resulting from the spy agency's drone strikes in Pakistan, Yemen, and Somalia. But here, at her very doorstep, were all manner of flying wings and whirlybirds carrying miniature, multispectral, gimbaled, inertially stabilized cameras. Flown from terra firma using handheld controllers and joysticks, various models streamed high-resolution still, video, and thermal imagery to displays on the ground. "What kind of camera was mounted on it? What kind of microphone?" Feinstein implored. "Could an enterprising person have fastened a firearm to it? These are questions that demand answers. Even with civilian drone technology in its infancy, privacy concerns are significant. I believe we should take these very seriously and move with steps to protect the privacy and safety of our fellow citizens before these capabilities are developed and unleashed, not after."

Yet as the nation reeled toward a robotic future, there was scant information available about domestic drone users and their rules of engagement. While CREW and Kentucky Senator Rand Paul helped part the FBI's veil of secrecy, the Electronic Frontier Foundation, a San Francisco-based nonprofit organization "dedicated to protecting privacy and civil liberties in the digital age," did yeoman work identifying the breadth of different users the FAA had authorized to fly unmanned aircraft. That took two rounds of legal action to bring to light the volume of information the agency had collected in support of those authorizations, recalled senior staff attorney Jennifer Lynch, who spearheaded the effort.

Media reports of a rising tide of domestic drones in 2010 and 2011 attracted the attention of the activist organization, which had libertarian roots. In a January 2011 article, *The Washington Post* described the Texas Department of Public Safety's use of a Wasp micro air vehicle to surveil a drug suspect two years earlier outside

The San Francisco headquarters of the Electronic Frontier Foundation. The nonprofit legal organization filed Freedom of Information Act lawsuits that compelled the FAA and the Department of Homeland Security to release their records on drones, revealing widespread use. *Courtesy of Dave Maass, Electronic Frontier Foundation*

Austin—the department's first public acknowledgment of the operation. A joint development of AeroVironment and the Defense Advanced Research Projects Agency, the Wasp weighed less than one pound and could be tossed into the air like a paper airplane. Nevertheless, it carried sophisticated, miniaturized electro-optical and infrared imaging cameras. "[W]hen drones come to perch in numbers over American communities, they will drive fresh debates about the boundaries of privacy," the *Post* predicted.[2] Lynch's recollection was that "it seemed to be a trend that was increasing or that had the potential to increase and something that at the EFF we hadn't really heard much about. We focus on police surveillance technologies and technologies used for national security, so we were especially interested in what was going on with drones."

The EFF faxed a Freedom of Information Act (FOIA) request to the FAA on April 11, 2011, seeking records on active and expired COAs and special airworthiness certificates the agency had issued for public and private entities to fly unmanned aircraft, as well as applications it had denied. The FAA acknowledged receiving the FOIA request four days later. "At the time I'd never seen what these [documents] even look like so I didn't know what to expect. It was sort of a shot in the dark," Lynch said. Nearly nine months later, when the FAA had failed to produce the requested records, the EFF filed suit in the US District Court for the Northern District of California in January 2012, requesting the court order the agency to immediately disclose the information.

The lawsuit and increasing media coverage appeared to motivate the FAA, which seemed at first to be caught unawares of the public interest in drones, Lynch said. At the time its UAS division was small, and documentation the agency collected was not organized in a format that could be released to the public. Lynch met in Washington, DC, with staff from the Department of Transportation's FOIA office, DOT, and FAA lawyers, and the person who was responsible for records at the FAA to devise a workable format for releasing the materials. In April 2012, the FAA produced two lists: one of the public entities that had applied for COAs and whether the authorizations had been granted or denied; and a second of manufacturers that had been granted special airworthiness certificates (SACs) in the experimental category to test fly their models. The EFF posted both lists to its website. The first revealed sixty-one entities had received COAs and forty-one were active users. The FAA had disapproved four applications. In addition to the military services and DARPA, several government agencies and the FBI had authorizations, as did police departments in Arlington, Texas; North Little Rock, Arkansas; Gadsden, Alabama; Ogden, Utah; Seattle, Washington; the Mesa County, Colorado, Sheriff's Office; the Polk County, Florida, Sheriff's Office; the Miami-Dade Police Department; and the city of Herington, Kansas. The EFF declared in a press release:

> These lists show for the first time who is authorized to fly drones in the United States.... Some of the entities on the COA list are unsurprising. For example, journalists have reported that Customs and Border Protection uses Predator drones to patrol the borders. It is also well known that DARPA and other branches of the military are authorized to fly drones in the US. However, this is the first time we have seen the broad and varied list of other authorized organizations, including universities, police departments and small towns and counties across the United States.[3]

The EFF found the list of thirty-six SACs to be less eye-opening, predictably consisting of aerospace manufacturers testing various models. Eighteen of those certificates remained active, covering the Raytheon Cobra, General Atomics Predator B, L-3 Viking 100, Honeywell T-Hawk, and Aurora Flight Sciences Goldeneye 50, among others.

The FAA's initial disclosure left questions unanswered. The COA list identified public entities that were authorized to fly drones, but not the models or the number of drones they flew. The FAA informed Lynch it had issued about 700 to 750 authorizations since 2006, and that some 300 were active—meaning some or all of the sixty-one public entities it listed had multiple authorizations. In July 2012, the agency began to release the actual COA application files on a rolling basis. "That's when we started to realize just how much information we were dealing with," Lynch said. "We just got thousands of pages of records. Most of the unredacted records came from universities and law enforcement agencies and some from the federal government. We saw a significant amount of unredacted records from the Army and the Air Force." The EFF also sought information from law enforcement agencies, including the Texas Department of Public Safety, the Miami-Dade police, and the Mesa County Sheriff, which were much quicker to respond and produced more information than the FAA, Lynch said. It partnered with MuckRock, a "collaborative news site" dedicated to helping people file public records requests, to conduct a "drone census" of every law enforcement agency with an FAA authorization to operate unmanned aircraft.

Congressional hearings during the summer of 2012 brought forth information that more than the sixty-one public entities the FAA originally identified had applied for COAs. Lynch said she sent a second FOIA request in August to gather material that had accrued since the EFF filed its first request more than a year earlier. In October 2012, the organization filed a second FOIA lawsuit against the FAA to obtain records produced since its first lawsuit. The agency was not necessarily dragging its feet; it was apparently overwhelmed and restrained in releasing information by other agencies within the federal government, Lynch believed. She argued:

> I don't think that the public should have to file a FOIA request and a lawsuit to get access to this type of information. However, having seen the quantity of records that we've received in response to the two FOIA requests, I have a lot of sympathy for the agency in trying to find a way to manage this kind of documentation. I believe that they are getting pushback from other agencies within the federal government on what they can and cannot release…I think that the other agencies should be producing records that can be released to the public. I think we have a difference of opinion on what should be released or not released.

Simultaneous with bringing a second legal action against the FAA, the EFF filed suit against the Department of Homeland Security in October 2012, seeking information about its practice of sharing Predator Bs; its Customs and Border Protection agency operated with other federal and state agencies, which was revealed by news accounts and the department's own inspector general. In a report earlier that year, the IG's office said the CBP had flown missions on behalf of other DHS agencies, the Bureau of Land Management, the FBI, the Department of Defense, the Texas Rangers, the

US Forest Service, and the National Oceanic and Atmospheric Administration. Among its findings: the customs agency "had not adequately planned for resources" to support the Predator fleet and had not adequately developed processes for accepting mission requests from other agencies, prioritizing those requests, and obtaining reimbursement for the missions flown.[4]

Prior to bringing the lawsuit, the EFF had filed a FOIA request in June 2012, which the DHS acknowledged receiving that July. In its reply, the department said the EFF "had not presented a convincing argument" that would entitle it to a blanket waiver of search and duplication fees for the requested records, so it would charge the foundation under regulations applying to representatives of the media.[5] After that, the EFF heard nothing more until after it filed the lawsuit that fall.

In the summer of 2013, the DHS, under then-secretary Janet Napolitano, delivered three years of flight logs, a 2010 "concept of operations" (Conops) document the department had submitted to members of the House and Senate appropriations committees, and other records in response to the FOIA suit. The flight logs—largely redacted of agency names, flight dates, and locations—indicated an eight-fold increase in the number of Predator missions the CBP flew for other agencies, from thirty in 2010 to more than 250 in 2012. These included missions in support of drug investigations, missing person searches, and fishing violations on behalf of state and federal law enforcement agencies, as well as aerial imaging missions supporting agencies including the US Geological Survey, the US Forest Service, and the Federal Emergency Management Agency, according to the EFF.[6] Subsequent to its initial release of records, the DHS produced further flight logs showing the CPB had flown 200 more flights for other agencies—700 as opposed to 500 over three years—than previously indicated.[7] In the 107-page Conops document, the department stated Predators would be used:

> In direct support of DHS operations...and in coordinated operations with other federal, state, local and tribal agencies. [M]ission payload data collected by the UA platform's sensors will be downlinked and received at compatible teleports...The UAS mission crew and/or imagery analysts at the AMOC (the CBP Air and Marine Operations Center) will review collected sensor data and conduct first-phase analysis. Following first-phase analysis, correlated sensor data and information will be disseminated to the tactical users and/or the supported command for that mission.[8]

Ultimately, the FOIA case the EFF mounted against the DHS was easier than its pursuit of information from the FAA because it involved a smaller quantity of records, Lynch said. "I think we did get some really good material in that case," she recalled. "We got most of a report on Predator drone flights and we got quite a bit of information about the times that CBP has loaned out its drones to other agencies."

As the EFF began to unravel the FAA's mountain of documents on drone flight authorizations, the unmanned aircraft industry and users began to self regulate. In July 2012, the AUVSI issued an industry "code of conduct," calling on its members to "respect the privacy of individuals" and "the concerns of the public," among other guidelines. "The future of UAS will be linked to the responsible and safe use of these systems," the code stated. "Our industry has an obligation to conduct our operations in a safe manner that minimizes risk and instills confidence in our systems," entailing both safe and "non-intrusive" operations.[9] In August 2012, the International Association of Chiefs of Police issued recommended guidelines for law enforcement agencies. The document advised local police to first assess the costs and benefits of using drones and then engage the community, civil liberties advocates, and the news media early in the planning process. In cases where police had grounds to believe that an unmanned aircraft would collect evidence of criminal wrongdoing and would "intrude upon reasonable expectations of privacy," they should obtain a search warrant in advance. Unless the imagery collected by an unmanned aircraft was required as evidence or as part of an ongoing investigation, it should not be retained by the agency.[10]

What the police did with the information their flying machines collected—not the flying machines—was the real issue when it came to privacy, AUVSI's Michael Toscano argued during a meeting in his office. He retraced the steps he took to testify before Senator Patrick Leahy's Judiciary Committee—how he drove in a car equipped with GPS, twice used his Bluetooth wireless device, passed through traffic lights fitted with cameras and pulled into a parking lot with still more cameras. "I would contend that the core issue is 'big data.' It's the collection, the analysis, the storage, the dissemination and the destruction of data," Toscano said. "But that's true with a lot of technologies that we have, whether it be the Internet, whether it be speed cameras, whether it be GPS. There's a lot of technology out there that does impede or affect our privacy."

A Coming "Surveillance Society"

On the issue of domestic surveillance by drones, if not by the National Security Agency scouring the nation's telephone records, Senator Feinstein agreed with even the American Civil Liberties Union, the powerful legal champion of individual rights and liberties. In testimony the ACLU's legislative counsel, Christopher Calabrese, submitted at the same Senate Transportation committee hearing that Feinstein addressed as a special witness, the organization warned of a coming "surveillance society in which every move is monitored, tracked, recorded and scrutinized by authorities."[11] Calabrese noted that the FAA Modernization and Reform Act of 2012 "dramatically accelerated the deployment of drones and pushed this issue to the forefront…The prospect of cheap, small, portable flying surveillance platforms threatens to eradicate existing practical limits on aerial monitoring and allow for pervasive surveillance. Our privacy laws are not strong enough to ensure that this new technology will be used responsibly and consistently with constitutional protections."

In a 2011 *Stanford Law Review* article, law professor Ryan Calo suggested another possible outcome. Rather than eroding privacy protections, he argued, ubiquitous, "disquieting" drones could help restore the public's "mental model" of what constituted a privacy harm, an awareness that was becoming clouded in the digital age of computers, the Internet, GPS, and other technologies. "They could be just the visceral jolt society needs to drag privacy law into the twenty-first century," Calo wrote. "…They represent the cold, technological embodiment of observation. Unlike, say, NSA network surveillance or commercial data brokerage, government or industry surveillance of the populace with drones would be visible and highly salient. People would *feel* observed, regardless of how or whether the information was actually used. The resulting backlash could force us to reexamine not merely the use of drones to observe, but the doctrines that today permit this use."[12]

Other organizations were concerned about drones and privacy. Ten days after President Obama signed the FAA reauthorization act in 2012, the Washington, DC based Electronic Privacy Information Center (EPIC) submitted a petition to FAA Administrator Michael Huerta that was ultimately signed by thirty-five consumer rights, human rights, technology and civil rights organizations, and 119 individuals, including famed consumer advocate Ralph Nader. The petition called on the agency to conduct a formal rulemaking process with the attendant notification and comment periods "to address the threat to privacy and civil liberties" resulting from increased civil and government use of "aerial drones." Unmanned aircraft fitted with gigapixel resolution cameras, infrared cameras, movement detection sensors, license plate readers, and facial recognition technology made pervasive domestic surveillance from above a real possibility, EPIC argued.[13] At issue: the Fourth Amendment protection against the government conducting "unreasonable searches and seizures." But overregulating drones also ran the risk of chilling freedom of speech as exercised by the public and the news media—something the First Amendment protects.

At that time there were no Supreme Court cases ruling on drones, although the court had allowed some warrantless surveillance from manned aircraft, according to the ACLU. In a May 1986 decision *California v. Ciraolo*, the court ruled that Santa Clara, California, police had not violated the Fourth Amendment by using a plane to spot marijuana growing in Dante Ciraolo's backyard, which was shielded from view at ground level by two fences.[14] The same year, the court ruled in *Dow Chemical Co. v.*

The Canadian-made Aeryon SkyRanger featured "hot swappable" modular payloads, including this sensor ball with dual electro-optical/infrared cameras, and flew up to fifty minutes on a lithium polymer battery. The SkyRanger was popular with police agencies, among them the Michigan State Police and the Royal Canadian Mounted Police. *Author photo*.

United States that the Environmental Protection Agency did not require a warrant to conduct an aerial inspection of Dow's facilities in Midland, Michigan.[15] In *Florida v. Riley*—a case decided in January 1989—the court found that Riley did not have a reasonable expectation of privacy when an officer with the Pasco County Sheriff's Office flew over his property in navigable airspace in a helicopter and spotted marijuana growing through missing panels on a greenhouse roof. In a dissenting option, Justice William Brennan made a prescient observation. "Imagine a helicopter capable of hovering just above an enclosed courtyard or patio without generating any noise, wind, or dust at all—and, for good measure, without posing any threat of injury," Brennan wrote. "Suppose the police employed this miraculous tool to discover not only what crops people were growing in their greenhouses, but also what books they were reading and who their dinner guests were."[16]

More recent decisions touched on component technologies relevant to unmanned aircraft and favored those being surveiled. In *Kyllo v. U.S.* in June 2001, the court held that federal Bureau of Land Management agents in Oregon had conducted a search within the meaning of the Fourth Amendment when they relied on a thermal imaging camera to detect heat lamps cultivating marijuana in Danny Lee Kyllo's garage. Since there was no advance warrant, the court reversed Kyllo's conviction on drug charges.[17] Similarly, attaching a GPS device to a vehicle to track its movements constitutes a search, the court decided in *U.S. v. Jones* in January 2012.[18]

Meanwhile, federal courts had relied on free speech analysis in holding that taking photographs of things that are plainly visible from public spaces is protected by the First Amendment, the ACLU said. "This right adheres regardless of whether the photographer is a member of the traditional media, and we believe the growth of citizen journalists and maturation of photographic technologies require strict First Amendment protections for all photographers, be they reporters, concerned citizens, protesters or artists," Calabrese stated in testimony. That would include Timothy Pool, who live-streamed video from a $300 smartphone-controlled Parrot AR.Drone during the "Occupy Wall Street" protests in 2011.[19] "The right of citizens to record the police is a critical check and balance," the ACLU said. "It creates an independent record of what took place in a particular incident, free from accusations of bias, lying or faulty memory. Visual evidence of police activity has often been crucial in investigating and reigning in police misconduct."

The media hoisted the First Amendment flag in May 2014, in an amicus brief supporting Raphael Pirker's challenge of the FAA's $10,000 civil penalty then being adjudicated by the NTSB board. Joined by household names including the Associated Press, Hearst Corporation, Cox Media Group, *The Washington Post*, and *The New York Times*, among other organizations, the media argued that the FAA's patchwork of regulatory and policy statements and "ad hoc" enforcement actions for what it deemed unauthorized drone operations chilled free speech. The *amici* took particular umbrage at the agency comparing the pursuit of the news to a pure business activity. "The FAA's position is untenable as it rests on a fundamental misunderstanding about journalism. Newsgathering is not a 'business purpose,' it is a First Amendment right," the media huffed.[20]

As a constitutionally protected activity, newsgathering using unmanned aircraft "unquestionably" should receive greater protection from government interference than afforded hobbyists and commercial users, the brief stated. For its part, the public stood to "benefit enormously" from the media's use of drones, which was something entirely different than law enforcement's use of unmanned aircraft. Any privacy issues arising from the media's drones could be addressed by existing state tort laws and statutes that provided sufficient remedies to address alleged invasions or privacy, the brief asserted. Indeed, tort laws—civil laws that sought to redress a wrongful act that either intentionally or through negligence caused harm to another person—appeared to protect the individual privacy rights drones threatened. The most relevant related to the harm caused by "intrusion upon seclusion," or intruding upon an individual's private affairs, according to the ACLU. State and federal Peeping Tom laws that imposed criminal penalties for viewing or photographing someone who is undressed or partially dressed when that person should reasonably expect privacy also offered a remedy to the illicit use of drones.

In February 2013, the city council of Charlottesville, Virginia, adopted a resolution drafted by The Rutherford Institute—a nonprofit civil liberties organization based in that city—urging Virginia's General Assembly to prevent police agencies from using UAVs equipped with "anti-personnel devices" and to prohibit government use of data collected from drones in state and federal court cases. The resolution also imposed a two-year moratorium on the use of drones within city limits. Charlottesville—home to the same University of Virginia campus that Raphael Pirker allegedly strafed with his foam flying wing—was the first city in the nation to limit the use of "police spy drones," according to The Rutherford Institute.[21]

In 2013, the Aerospace States Association, the Council of State Governments, and the National Conference of State Legislatures produced a set of recommendations for state policy makers wrestling with the onset of drones. Foremost among them: states should require warrants for government surveillance of individuals or their property using unmanned aircraft, and should consider prohibiting commercial drone tracking of individuals without their consent.[22] The ACLU tracked what it called "privacy protective" legislation proposed and enacted by state legislatures. According to its research, in April 2013, Idaho Governor C.L. "Butch" Otter signed into law the first bill in the nation prohibiting police from using drones to conduct investigations without a probable cause warrant. The law also placed restrictions on the private use of drones, prohibiting photography of private property without the owner's written permission.[23] That year—the first full year since the president signed the 2012 FAA Modernization and Reform Act—forty-three states took up drone measures and eight enacted legislation. By the spring of 2014, twelve states had enacted legislation. "Like last year, almost all of the bills we're seeing require law enforcement to get a probable cause warrant before using a drone in an investigation," the ACLU said. "This year, though, much of the proposed legislation is more complex, taking into account the more challenging issues like what to do with information that is collected incidentally to lawful drone use, how long law enforcement can hang on to drone-collected data, and how to handle government access to information collected by third-party drones."[24]

Douglas Marshall, a former aviation lawyer and professor, advised attendees at AUVSI's Unmanned Systems conference in 2014 that freedom of speech would likely drive the first drone-specific case elevated to the nation's highest court. "There will be a UAS case that will end up in front of the US Supreme Court eventually. It has to happen," said Marshall, then with the UAS Technical Analysis and Applications Center

Physical Sciences Inc., of Andover, Massachusetts, developed the InstantEye quadcopter. Weighing less than two pounds, the craft carried an infrared illuminator and three cameras on its belly. Another possible payload was a detector package for radiological or nuclear threats. *Author photo*.

at New Mexico State University. "This industry is too big, it's growing too fast and there's too much going on for there not to be. I don't think it's going to be a criminal Fourth Amendment prosecution case. I think it's going to be a First Amendment case driven by the media claiming that they ought to be able to use these things...unfettered by any kind of federal aviation regulations."

Yet it fell to the FAA, the agency whose mantra was safety first, to develop a drone privacy policy. In a special notice dated February 14, 2013, the FAA announced it would insert privacy requirements in the "other transaction agreements" it planned to sign with public entities chosen to operate the six UAS test sites Congress required. While this was not the full-fledged rulemaking that EPIC sought in its petition, the agency would field comments on the privacy language concurrently with soliciting proposals from test site applicants. "Test site operators will be required to establish a privacy policy that is public and builds confidence and trust," the notice stated. "Furthermore, the FAA expects that the information gathered about UAS operations at the test sites will contribute to the dialogue among privacy advocates, policymakers and the industry about how to address broader questions relative to the technologies used."[25] The test sites were not just testing unmanned aircraft; they were floating a privacy policy.

A week after serving notice, the FAA published its proposed privacy requirements for comment in the *Federal Register*. In addition, the agency held a "public engagement session" by teleconference on April 3, 2013, allowing each speaker three minutes to comment. Prefacing the event, Jim Williams, manager of the UAS Integration Office, said the FAA was prepared to begin the test site selection process the previous summer, but held back to address privacy concerns.[26]

The FAA released the privacy policy—signed by its acting chief counsel, Marc Warren—simultaneously with the first UAS road map on November 7, 2013. It reported receiving ninety-nine written and fifty-three spoken comments to the public notice and listening session, respectively. They ranged from commenters who believed the FAA should focus on safety and not engage in regulating privacy to those who felt the agency should punish privacy violations. The final conditions contained in the OTA agreements with test site operators required each site to have a privacy policy and make it publicly available; to have a mechanism to receive and consider public comments; to use as guidance "Fair Information Practice Principles" regarding the privacy of information about individuals; to comply with all applicable federal, state, and local privacy laws; to maintain a record of all UAS operating at the site; to require each UAS operator to have a written plan for the use and retention of data collected by the aircraft; and to conduct an annual review to verify compliance with the policy. The FAA reserved the right to suspend, terminate, or amend the OTA agreements based on privacy violations flagged by federal or state law enforcement authorities, or on changes to applicable law.[27]

The FAA's privacy policy met with a mildly positive reception from civil liberties organizations. In general, liberties groups saw the policy as a step in the right direction toward protecting privacy rights that did not go far enough. "We were happy that the FAA took at least some stand on privacy. I think that the agency is taking a cautious approach by just including privacy regulations in these privacy policies at the six test sites when the agency could have gone farther and required, for example, all COA applicants to provide information on how they're going to protect privacy," the EFF's Jennifer Lynch remarked. In testimony to the Senate Transportation Committee, the ACLU's Christopher Calabrese said: "We...applaud the FAA for beginning the process of exploring privacy controls and its continuing commitment to using the Fair Information Practice Principles as an appropriate framework (for) making those determinations."

Nevertheless, the ACLU called on Congress to enact statutory protections that would reinforce the Fourth Amendment protection against unreasonable searches and seizures. It recommended a two-pronged approach differentiating between government and private sector use of drones. "Congress should place immediate, robust restriction on the government use of drones, especially as part of criminal investigations, in order to prevent mass aerial surveillance," the organization stated. Use of drones by the private sector "also presents serious privacy risks, though those risks must be counterbalanced by real and important First Amendment values." Among "general propositions," Congress should take care to regulate the improper use of photographs—not the actual expression—and in no case single out newsgathering for special restrictions. "The constitutional right to photograph anything visible from a public vantage point—including, in particular, government activity—must be protected." [28]

In a blog post, Gautam Hans of the Center for Democracy & Technology said the FAA's policy was "strong" in some ways. "The FAA has recognized that it has the institutional capacity to protect privacy, resisting claims that it doesn't have the expertise to do so," he wrote. The policy required drone operators to document their use and retention of any data collected. It required test site overseers to audit the operations for compliance with privacy rules and make public the information. Ultimately, though, the center was disappointed. "Taken together, the final privacy requirements may create an inconsistent patchwork of protection across the test sites depending on the specific desires of each location," Hans said. "Given that the FAA has jurisdiction over the entire national airspace, it would have been more appropriate for the agency to mandate consistent test-site privacy standards, rather than allow for conflicts to develop as part of the program."[29] Legislation that Massachusetts Senator Edward Markey proposed ("The Drone Aircraft Privacy and Transparency Act") would update the requirements Congress imposed in the 2012 FAA reauthorization act and address privacy in a uniform way, he said.

A graying liberal Democrat from working-class Malden, north of Boston, Markey had served thirty-seven years in the House, where he first introduced the drone privacy legislation. After President Obama tapped Massachusetts Senator John Kerry as his secretary of state to replace the retiring Hillary Clinton, Markey won a special election in June 2013 to serve out the eighteen months remaining in Kerry's Senate term. From that loftier perch, he reintroduced the drone privacy bill in November 2013.

Markey also sat on the Senate Commerce, Science, and Transportation Committee. In that capacity, at the same hearing that Feinstein addressed in January 2014, he served notice he was committed to shepherding the legislation to the President's desk. "There are benefits to this technology…and that's great," he said. "It's an industry; we should be in the industry. But while there are benefits to drone use, there are also risks of misuse. These 20th century 'eyes in the sky' shouldn't become 'spies in the sky.' Just as there are rules of the road, there have to be rules for the skies if we're going to commercialize." New Jersey Senator Cory Booker then handed Markey what appeared to be an orange-and-blue painted AR.Drone quadcopter. "This drone here has two—count them, two—independent cameras, and they can be purchased online for under a hundred dollars," Markey declared. "With an iPhone app you can fly this over the Capitol right now, or over anyone's backyard."

Markey turned to the ACLU's Calabrese, who was seated at the witness table, and reeled off a series of questions without waiting for answers:

> If a company decides to fly a drone over my backyard and video me, would there be any way for me to know it? If I look up and see a drone flying over my house, under the FAA's current plan, is there any way I can find out what information that drone is collecting? If that drone happened to take images of me in my backyard, are there any federal laws requiring them to delete those pictures? Would I be able to find out how a company uses those pictures or any data they collected of my family or anyone else's family? Would I at least be able to find out who owns or is operating that drone?

In the penultimate year before the FAA's mandate to open the airspace to drones, many such questions remained.

In the meantime, Congress exerted pressure on the agency through an appropriations bill—legislation that released the money government agencies were authorized to spend. In an "explanatory statement" accompanying the 2014 Transportation, Housing and Urban Development, and Related Agencies appropriations act, Congress directed the FAA to conduct a study on the implications of introducing unmanned aircraft into the airspace on individual privacy. While acknowledging the agency's primary mission was to protect civil aviation safety and to provide an efficient airspace system, "the FAA also has a responsibility to provide the Congress with information and analysis on civil aviation issues," the appropriators said. "[W]ithout adequate safeguards, expanded use of UAS and their integration into the national airspace raise a host of concerns with respect to the privacy of individuals." The required study, due within eighteen months of the law's enactment (mid-2015), "should address the application of existing privacy law to UAS integration; identify gaps in existing law, especially with regard to the use and retention of personally identifiable information and imagery; and recommend next steps" the agency could take to address the impact on privacy from the widespread use of drones.[30]

The White House also took notice. *Politico* reported in July 2014 that President Obama planned to issue an executive order to the National Telecommunications and Information Administration to develop drone privacy guidelines. An Obama administration spokesman told reporters that a government-wide, interagency effort was underway to coordinate federal policy on the issue.[31] On February 15, 2015, the White House issued a presidential memorandum on the federal government's use of unmanned aircraft on the home front, directing federal agencies that used drones—a group that included the FBI and the Department of Homeland Security—to honor established privacy protections, regularly update their policies and procedures for the collection, retention, and dissemination of information obtained by unmanned aircraft, improve their oversight of operations, and provide public notice when they used drones.

CHAPTER 8

TECHNOLOGY CHALLENGES

All the FAA asked was that an unmanned aircraft fly like one with a pilot on board. UAVs should comply with the same rules as manned aircraft without crowding out existing users of the airspace, forcing them to change their operations or degrading safety. They should not endanger people or property on the ground. They should be designed, manufactured, and maintained to comparable standards. The same held true for the people who operated UAVs. Pilots, visual observers, launch and recovery specialists, and maintenance personnel were to be comparably trained and certified, as were the instructors, inspectors, and controllers overseeing them.

The goals and expectations for introducing UAVs into wider, civilian skies were simply stated, but as always the devil was in the details. The FAA managed the world's busiest and most complex airspace system. FAA operated and contract air traffic control towers handled fifty million takeoffs and landings per year.[1] There were 70,000 flights each day, moving two million passengers. At any given moment, there were 5,000 planes in the sky.[2] The system hosted airliners, corporate jets, general aviation turboprops, piston-powered airplanes, military aircraft, and helicopters, all with different performance characteristics. The general aviation community in the United States dwarfed that of any other nation, with a fleet consisting of more than 220,000 aircraft.

Without a man or woman at the yoke or sidestick controller in an aircraft, a UAV was challenged to maintain "self-separation" from other aircraft and in the worst case scenario dodge an airborne collision; requirements of federal aviation regulations Parts 91.111 and 91.113. The first stated in part that: "No person may operate an aircraft so close to another aircraft as to create a collision hazard." The second established right-of-way rules, requiring in general that pilots remain vigilant to "see and avoid" other nearby aircraft. With UAVs, these functions were performed by a pilot on the ground, who communicated by voice or data messages with an air traffic controller, or through automatic responses the aircraft made based on its sensor readings. The pilot needed a reliable command and communications (C2) radio link with the air vehicle and precise, uninterrupted information on its status.

Keeping UAVs out of harm's way was no trivial science project. Protecting passenger-carrying aircraft from midair collisions took more than three decades to master. The traffic alert and collision avoidance system (TCAS) that was a fixture in most large aircraft was borne of tragedy. The "seminal" event that spurred its development was the June 30, 1956, midair collision of a TWA Super Constellation with a United Airlines DC-7 over the Grand Canyon that killed 128 people on board the two aircraft.[3] In the aftermath of the crash, the airline industry pressured suppliers to develop an airborne collision avoidance system that would act as a safeguard of last resort when ground controllers failed to keep aircraft safely separated. Two further midair crashes involving airliners and military jets in 1958 motivated Congress and the administration of President Dwight Eisenhower to pass the Federal Aviation Act of 1958, creating the modern-day FAA.

But keeping aircraft flying in close proximity at high speeds from colliding with each other proved to be a monumental engineering challenge. Over decades, the solution evolved from systems based on radar and other "non-cooperative" sensors that independently detected nearby aircraft to a "cooperative" system, in which each aircraft must have a transponder. The transponder sends out radio signals, or "interrogations," to other transponder-equipped aircraft in the vicinity and calculates their relative position based on the signals it receives. Directed by Congress, the FAA issued a rule in 1989, requiring that all passenger-carrying aircraft with more than thirty seats be equipped with TCAS II systems capable of providing "resolution advisories"—recommended escape maneuvers—to pilots by 1991. Congress later amended the requirement, extending the equipment deadline to 1993. In 1995, the FAA mandated TCAS I with less sophisticated alerting capability for turbine-powered aircraft with ten to thirty-one seats.

In their 2012 book *Flying Safer Skies With Electronic Eyes*, former Bendix Radio engineers Jack Shagena and James Dalmas recounted the many twists and turns that led to the development of TCAS. "This is an aviation story not so much about pilots and entrepreneurs, but about engineers and scientists and programmers at companies past and present," this author wrote in the introduction to that book. The TCAS story had "multiple dimensions of money, politics and the sometimes conflicting priorities of manufacturers, the FAA, Congress, the airline industry, pilots and air traffic controllers. Ultimately, though, it is about the triumph of technology for the public good."[4] Easing UAVs into the airspace would require a similar triumph of technology; whether for the public good was debatable. Some of the same organizations that contributed to the development of TCAS—the FAA, NASA, the MITRE Corporation, MIT Lincoln Laboratory, RTCA, and leading manufacturers—figured prominently when it came to unmanned aircraft. While the mainstream media hyperventilated over drones darkening the skies, some of the best technical minds in the United States worked the technical challenges of flying UAVs safely with other air traffic.

Congress established MITRE in 1958 as a private, nonprofit corporation to provide technical guidance to the federal government, initially for military purposes. With principal locations in Bedford, Massachusetts, and McLean, Virginia, it managed a portfolio of federally funded research and development centers. In November 2004, the MITRE Center for Advanced Aviation System Development produced a paper entitled "Issues Concerning Integration of Unmanned Aerial Vehicles in Civil Airspace" containing a laundry list of stumbling blocks the unmanned aircraft community faced gaining routine and safe entry into non-restricted, civilian airspace. Technological hurdles included the absence of certification standards for UAV structures and systems, their operators, and the types of operations they could perform; and the lack of an affordable collision avoidance system and protected radio frequency spectrum to control them. Among less tangible, though no less daunting, hurdles were "public apprehension" and acceptance of UAVs, resistance from current airspace users, and the lack of a "business case" that would make flying them financially worthwhile. "Integration of UAVs into civil airspace and their potential market success depends on a complex set of technical, economic, political and legal factors," the paper stated. "Unlike the early years of aviation, UAVs do not operate in empty skies. Rather they must contend with a mature civil aviation system—one filled with aircraft, controlled and monitored by complex systems, dominated by large commercial markets, saturated by interest groups and governed by a voluminous regulatory structure."[5]

Around the same time, MITRE became involved in NASA's Access 5 program with the UNITE aerospace industry group, which was focused on achieving routine access to the national airspace system within five years for a high-altitude, long-endurance unmanned aircraft. MITRE also provided technical assistance to the Department of Homeland Security, which in 2004 awarded a lease to Israel's Elbit Systems and its EFW subsidiary in Fort Worth, Texas, to fly the Hermes 450 along Arizona's southern border as part of the Arizona Border Control Initiative. "We're focused on our sponsors' missions," explained Andrew Lacher, MITRE's UAS integration research strategist. "We listen to our sponsors and we use that to anticipate what they might be needing in the future. We got people together from across the company to (ask): what are the FAA, what are the DOD, what are the DHS objectives and what are the challenges from a technical, operational and policy perspective?"

Airborne Sense and Avoid

Proven effective and venerable though it may have been, the FAA did not consider the TCAS system alone capable of preventing UAVs from colliding with other objects. At AUVSI's Unmanned Systems conference in 2011, Steve George, an aerospace engineer with the FAA's unmanned aircraft program office—the precursor of the UAS Integration Office—shared the findings of an analysis the agency conducted using TCAS on an unmanned aircraft. For a UAV pilot to rely on a TCAS traffic display on the ground "is unsafe in class A, B, C, D, E and G airspace," George said. "The representation of the

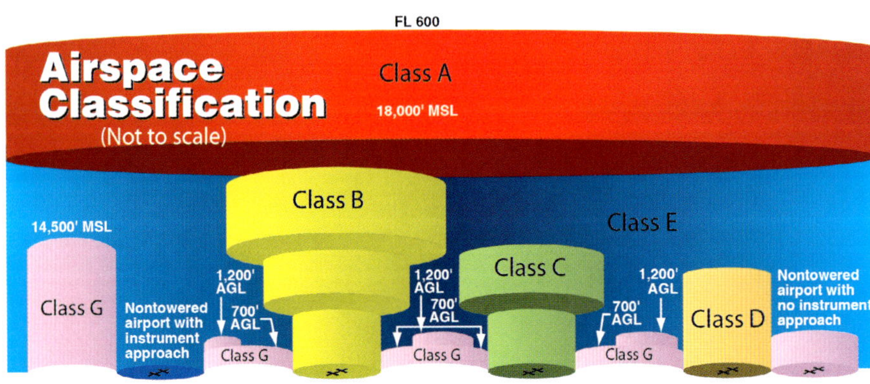

This graphic depicts the US airspace classification system. The FAA disallowed flying unmanned aircraft in Class B airspace surrounding the nation's busiest airports. Technology efforts were concentrated on flying large UAVs through lower altitudes to Class A airspace above 18,000 feet MSL (mean sea level). AGL is above ground level. *Courtesy of the FAA*

traffic situation is inaccurate, potentially incomplete and misleading. It's unsafe to be used as the sole basis for maneuvering."[6] TCAS was designed for use on manned aircraft and was not compatible with UAVs. There was the potential for excessive delay or "latency" of commands transmitted over the C2 link or of losing the link altogether, preventing the ground crew from executing a timely evasive maneuver. TCAS relied on other nearby aircraft having transponders and could not detect non-cooperative aircraft that did not. It was designed for specific aircraft performance characteristics—including climb rate—that some UAVs could not match. Its use by a multitude of UAVs could saturate the already congested 1090 MHz radio frequency band on which TCAS and other systems depended.[7]

Another technology had evolved that complemented TCAS in protecting aircraft from midair collisions. Automatic Dependent Surveillance-Broadcast (ADS-B) was a concept for tracking aircraft the FAA advanced in a July 1996 document called "Surveillance Vision Plan, Revision 2." Through ADS-B, a transponder-equipped aircraft continuously broadcasted its precise position in space—calculated by an onboard GPS receiver—as well as its identification and other information to radio stations on the ground and nearby ADS-B equipped aircraft. Like TCAS, it was an air-to-air method of surveillance, as opposed to depending on a fixed ground radar with a mechanically rotating antenna. The FAA's intent was to improve air traffic control and eventually replace secondary surveillance radars that interrogated and received signals from aircraft transponders with a system providing greater overall coverage, better accuracy, and higher update rates—but one that also held promise for airborne collision avoidance. ADS-B was the new surveillance paradigm the FAA chose to replace ground-based radar under its NextGen modernization effort. In August 2007, the agency awarded Exelis—then known as ITT—the contract to build the nationwide ADS-B network of ground radio stations. (Harris Corporation of Melbourne, Florida, acquired Exelis in May 2015.) In May 2010, the FAA issued a final rule requiring operators equip their aircraft for ADS-B "Out" position broadcasts by 2020. It deferred requiring ADS-B "In"—the capability to display the air traffic scene in aircraft cockpits—which required expensive display and software improvements.

Unmanned aircraft research efforts embraced the new technology. In 2011, MITRE joined with NASA Langley Research Center, the University of North Dakota, and Draper Laboratory to test a UAV collision avoidance system using ADS-B as the sensor. The objective of the Limited Deployment-Cooperative Airspace Project was to test a "cooperative autonomous sense-and-avoid" system. It was cooperative, in that it depended on transponder-equipped aircraft signaling by ADS-B; autonomous, in that it worked in the event the C2 link to the aircraft was interrupted; and sense-and-avoid, in that it would provide the capabilities of "self-separation" and possibly collision avoidance as well. "At the time we undertook the effort, there weren't a lot of people talking about ADS-B as a potential surveillance source," said Lacher. "A lot of people dismissed it immediately because not everyone is equipped with ADS-B."

The tests involved flying a NASA-Langley Cirrus SR22 single-engine airplane as a "surrogate" UAV with a safety pilot aboard to evaluate software maneuver algorithms based on ADS-B detection of nearby aircraft targets. The project ran some 400 encounters in flight against a UND "intruder" airplane over eastern North Dakota and against a NASA intruder over Tidewater, Virginia. The partners demonstrated the sense-and-avoid capability against a UND Cessna 172 during a flight test at Grand Forks International Airport on September 20, 2012. "More than 100 leaders from academia, industry, government, the military and the general aviation community came to the Grand Forks Airport to observe an LD-CAP flight demonstration," they announced in a press release. "Live, on a large screen, they saw how a remotely piloted aircraft…could detect a plane in its vicinity, then automatically be safely maneuvered away from any possible conflict."[8]

One outcome of the project for MITRE was that it developed an "algorithm evaluator" it applied in computer simulations that were capable of "literally millions of encounters" to determine if sense-and-avoid algorithms worked as expected. "We weren't evaluating the algorithms per se; we were demonstrating that it is feasible to use ADS-B to do detect and avoid, and we were using the algorithms to confirm the performance of the simulation," Lacher said. "We were never suggesting (ADS-B) was going to be adequate as the only surveillance source; we were just suggesting it could be a very good surveillance source. And more importantly, we were focused on how do you actually evaluate detect-and-avoid capabilities?"

In March 2012, NASA's Dryden Flight Research Center (renamed NASA Armstrong in 2014) flew its Ikhana UAV—a General Atomics' Predator B—for the first time with an ADS-B transponder. The FAA's William J. Hughes Technical Center, located at Atlantic City International Airport in New Jersey, recorded ADS-B signals from the nearly three-hour flight that NASA conducted over its Western Aeronautical Test Range in the Mojave Desert. The test also marked the first shakedown flight of hardware NASA was studying under its five-year "UAS Integration in the NAS" project, scheduled to run to 2015. The project's aim was to develop a "body of evidence" to prove requirements the FAA and standards organizations could use to qualify civil UAVs.[9] Manufacturer General Atomics tested ADS-B on a US Customs and Border Protection Guardian off the coast of Florida in August 2012.

That same summer, R³ Engineering—a small technology firm based in Lusby, Maryland—declared it had demonstrated the first "ADS-B-based, fully autonomous collision avoidance sequence" of an unmanned aircraft. The trial involved a research UAV the firm equipped with its All Weather Sense and Avoid System, an ADS-B radio developed with funding from the Office of Naval Research and manufactured by Appareo Systems of Fargo, North Dakota. R³ conducted the trial at a former US naval base in Argentia, Newfoundland; it included potential collisions between the UAV and another unmanned aircraft and with stationary and moving ground hazards. "ADS-B messages received and processed by the (system's) internal computer indicated that the aircraft was on a collision course with another ADS-B equipped target," the firm reported. The system "analyzed the flight tracking data; determined that a potential breach of the aircraft's 'collision volume' was imminent; determined an optimum collision avoidance maneuver; and sent commands directly to the autopilot." The autopilot "responded to the commands, turning the aircraft away from the intruder; and when clear of the conflict, returned the aircraft to its preprogrammed flight path."[10] In April 2013, R³ demonstrated autonomous collision avoidance maneuvers of two Tiger Shark UAVs in a Navy-funded project at Yuma Proving Ground, Arizona. The Tiger Sharks, manufactured by Navmar Applied Sciences Corporation, were fitted with Cloud Cap Technology Piccolo autopilots and R³ Engineering's avionics box, which weighed less than one pound.[11]

While ADS-B held promise for preventing midair collisions of unmanned and other aircraft, technologists did not consider it the Holy Grail for introducing UAVs en masse to the airspace system. Any future collision avoidance box would likely draw information from a mix of technologies.

The US Air Force Research Laboratory at Wright-Patterson Air Force Base near Dayton, Ohio, began looking into sense-and-avoid technologies for unmanned aircraft in 2001. The laboratory started acquiring systems in 2007, and demonstrating them in 2008. The AFRL's original focus was on TCAS and ADS-B, cooperative systems that depended on other nearby aircraft having transponders to broadcast and reciprocate their positions. "The notion was to provide some kind of integration function in a software algorithm that would allow us to integrate those two sensor pieces together into a single picture to predict when there might be an encounter and how to avoid an encounter between two aircraft," said Paul Schaeffer, the Air Force's Airborne Sense-and-Avoid program manager. "At that point in time, we still hadn't really come to grips with the terms 'self-separation' and 'collision avoidance.' It was simply a matter of trying to work for safety in the airspace." As the effort progressed, the AFRL brought in non-cooperative systems—radar and electro-optical (EO) cameras—recognizing that many smaller general aviation aircraft did not have transponders and therefore could not be "seen" by cooperative systems. It entered into contracts with Colorado Engineering, of Colorado Springs, Colorado, to develop an electronically scanned C-Band radar; and with Defense Research Associates of Beavercreek, Ohio, to develop an EO sense-and-avoid system.

The AFRL's "multiple intruder autonomous avoidance" series of flight tests, which involved Northrop Grumman, Calspan Flight Research, and the FAA Technical Center, demonstrated a multi-sensor solution that combined TCAS, ADS-B, EO cameras, and radar. The partners conducted flights from Calspan facilities at Niagara International Airport in New York, pitting a Calspan Learjet fitted with multiple sensors and acting as an unmanned surrogate against intruder aircraft flown by Calspan and the FAA. "Very often the FAA is actually providing the target aircraft along with the truth data of where those target aircraft are. They've been a huge help to us in the development of the technology," Schaeffer said.

Initial sense-and-avoid research focused on Northrop Grumman's high-flying RQ-4 Global Hawk surveillance UAV, which drove the AFRL's choice of a radar operating in the C-Band, Schaeffer said. "Of the sensors that we're putting onto the platforms, the radar is the one that is special-purpose-built for each of the platforms," he explained. "The things that are driving that are the size, weight and power requirements for the radars themselves on the aircraft, as well as frequency compatibility with all of the other electronic sensors that are on the aircraft. For the Global Hawk we have chosen a C-band radar because that is the band which provides the least inherent jamming of our own sensors." In another AFRL-sponsored effort, Sierra Nevada Corporation demonstrated an electronically steered X-Band radar for sense-and-avoid duty.[12]

In 2009, the Air Force and Navy combined their sense-and-avoid research efforts under the Broad Area Maritime Surveillance program, the Navy's requirement for a "persistent" maritime surveillance UAV. The previous year, the Navy had selected Northrop Grumman's RQ-4N Global Hawk as the "BAMS" platform, which it later christened the MQ-4C Triton. Northrop Grumman contracted ITT Exelis to supply the sense-and-avoid radar subsystem. Due to "funding constraints and program priorities" Schaeffer said, the Navy stopped funding the sense-and-avoid component in 2012, leaving the Air Force as the sole entity behind the program. The Navy's deputy program manager, Sean Burke, later briefed reporters on technical issues the service faced when it tried to build an active electronically scanned array (AESA)

The US Navy's MQ-4C Triton was parked in a hangar at the Patuxent River Naval Air Station in Maryland following its first cross-country flight in September 2014. The Navy planned to fly the Global Hawk derivative over international waters, but it was challenged to develop collision avoidance capability. *Author photo*.

radar into the Triton's nose to protect against collisions. "The sense-and-avoid technical solution turned out to be very challenging for us, and one of the keys to that challenge was trying to squeeze an AESA radar into the size (available), with the weight and power and cooling allocations that we were able to provide for that system," said Burke. "Miniaturizing that kind of a radar suite has turned out to be more challenging than we initially thought. So we're stepping back from it, we're taking a hard look at what the right way forward is."[13]

The Air Force began a formal sense-and-avoid system acquisition program in October 2012, for what at the time remained a Global Hawk specific technology, Schaeffer said. By then, the service also recognized other UAVs, such as General Atomics' Predator and its larger sibling, the MQ-9 Reaper, faced the same requirements if they were to fly in civilian airspace. So too did the Navy's Triton and the Army's MQ-1C Gray Eagle. "To better serve both the Department of Defense and the taxpayer (we decided) that we should build a system to be common across all the UAV platforms to minimize the amount of money that would be required to develop independent solutions for each UAV," he said.

In January 2013, the Air Force adopted a new strategy to develop a common airborne sense-and-avoid system and the ABSAA program thereafter became C-ABSAA. In September, the Air Force issued a request for information to industry seeking potential sources to develop a C-ABSAA system "for large remotely piloted aircraft systems including the Predator and Global Hawk." The system was to be "sensor agnostic"—capable of accepting data from multiple sensors—and capable of seeing both cooperative and non-cooperative targets. It was to use a display methodology common to different ground control stations. Given the state of the technology, the service said it expected to begin the program's engineering, manufacturing, and development phase in 2017.[14]

When the strategy shifted from a Global Hawk specific system to one that was "platform agnostic," the Air Combat Command, which set Air Force requirements, directed the C-ABSAA program to rewrite the sense-and-avoid system requirements it had thus far developed. "So we're in the process of rebuilding requirements for a common system, starting basically from scratch," Schaeffer related in early 2014. From a requirements standpoint that slowed progress toward building a system by about two years, he said. With the service predicting a contract award in 2017 to begin C-ABSAA development, Schaeffer said: "We would expect that we would field this system on a yet-to-be named platform in the mid-twenties." The Air Force had already invested some $150 million over thirteen years to advance airborne sense-and-avoid technology. It owned the data rights "for most everything that we have built," which it would make available to industry to build a system.

Collision Avoidance X Factor

While industry and government testers moved the needle on unmanned sense-and-avoid, the FAA was developing a new collision avoidance capability as part of its Airborne Collision Avoidance System X program. The agency envisioned ACAS X as a "family of systems," an evolution of TCAS II that would also accept ADS-B signals. The concept originated with Mykel Kochenderfer, then with MIT Lincoln Laboratory, who around 2008 convinced the laboratory to fund its development. The system was originally intended for unmanned aircraft, but the FAA's TCAS program office, under Neal Suchy, saw its potential to replace the existing TCAS system for manned aircraft, said Wesley Olson, a member of the laboratory's technical staff. Suchy promoted ACAS X to his bosses at the FAA, which in July 2013 committed to funding ACAS Xa, a TCAS replacement incorporating new threat logic. Another variant—ACAS Xo—was part of the development; it would be finetuned for specific airspace procedures that would cause ACAS Xa to trip false alarms, such as arrivals to closely spaced parallel runways. The FAA assembled a technical team to support the effort including Lincoln Laboratory, Johns Hopkins Applied Physics Laboratory, MITRE, and contractors Aurora Sciences and Regulus Group.

In August 2013, the FAA conducted ACAS Xa/Xo flight tests at the Hughes technical center. The agency flew some 120 encounter scenarios using a Convair 580 research airplane fitted with a prototype Honeywell traffic computer flying against a second Convair 580 and other aircraft and proved the new threat logic could work. "Based on the demonstrated performance, the current version of the system reduces the risk of a midair collision by almost 50 percent overall, and reduces the alerts by almost 50 percent as well," said Olson. He explained that ACAS X was designed to be flexible; it would accommodate a mix of sensors and adapt to a variety of aircraft. He said:

> The goal of the system is to employ more advanced computational techniques that explicitly consider uncertainties both in terms of how vehicles respond and also in terms of surveillance noise, and be somewhat sensor agnostic—that

> is to allow other than the active transponder-based surveillance. It could also be tuned for the performance of various other types of aircraft. The current TCAS assumes that vehicles will be able to climb and descend at least 1,500 feet per minute and that is not an appropriate assumption for many aircraft, rotorcraft, smaller GA aircraft [and] UAS."

Whereas the FAA formally advanced the ACAS Xa and Xo variants with the expectation that its advisory organization RTCA would complete operational performance standards for the new system in 2018, ACAS Xu for unmanned aircraft remained a research project. At the time, the FAA had no formal program to advance ACAS Xu as a collision avoidance capability, but it continued researching the system as a last-resort solution to preventing midair crashes and also by itself or with other systems to enable self-separation. Flight tests were planned in 2014 and 2015 using a system adapted to NASA's Ikhana. "ACAS Xu is going to provide the capability for vertical collision avoidance advisories that are specifically tuned for the performance of the Ikhana vehicle, which is typically less than 1,500 feet-per-minute climb and descent rate, as well as provide horizontal collision avoidance maneuvers against non-cooperative targets, using in all likelihood General Atomics' 'due regard' radar. That is the current plan," Olson said.

"Due regard," a term framed in the 1944 Chicago Convention that established the International Civil Aviation Organization, referred to ICAO's requirement that military and state-owned aircraft be flown with "due regard for the safety of navigation of civil aircraft." The US Department of Defense, in a policy directive called Instruction Number 4540.1, interpreted due regard as requiring aircraft engaged in non-routine, military contingencies over international waters "be equipped with airborne radar that is sufficient to provide separation between themselves, aircraft they may be controlling, and other aircraft." In those situations, an aircraft commander must "act as his or her own air traffic control and to separate his or her aircraft from all other traffic." The term later came to be used in reference to airborne sense-and-avoid radar.[15]

In November 2013, General Atomics conducted the first flight of a sense-and-avoid system with self-separation functionality on a company-owned Predator B at its Palmdale, California, flight operations facility. The system combined a BAE Systems' military Identification Friend or Foe transponder capable of receiving ADS-B signals, General Atomics' due regard radar, and a Honeywell TCAS computer processor working in unison to detect and track cooperative and non-cooperative aircraft. "This marks the first time the system has functioned as a true 'system of systems' to detect every class of aircraft equipage and paves the way for a due regard capability," General Atomics announced. "[T]he Honeywell sensor fusion algorithm excelled in combining the multiple sensors' outputs and fusing them into a single track picture to send to the ground."[16]

A year later, General Atomics, the FAA, NASA, and suppliers started flying a proof-of-concept collision avoidance system on NASA's Ikhana air vehicle, testing for the first time the FAA's ACAS Xu algorithm as well as a pre-production version of General Atomics' due regard radar. Following functional flight tests of the system on a Predator B from September 4 through 10 at Palmdale, the partners conducted the first flight on the Ikhana on November 18, 2014, at NASA Armstrong Flight Research Center in Edwards, California. The Ikhana—named after a Choctaw word meaning

The NASA Ikhana is an MQ-9 Predator B the space agency has deployed for unmanned aircraft collision avoidance research with the FAA and industry partners. Ikhana is a Choctaw word meaning "intelligent" or "aware." *Courtesy of NASA*

"intelligent" or "aware"—would serve as the primary test aircraft, flying against FAA and Honeywell target aircraft in the Edwards Air Force Base restricted area. General Atomics planned to complete the pre-production due regard radar by the first half of 2015; the full sense-and-avoid equipment suite would be designed to specifications the company expected RTCA to deliver in 2016. "Our goal as a company is to be basically the prototype system for the community," said Brandon Suarez, General Atomics' sense-and-avoid program lead.

Yet another approach to collision avoidance that MITRE was investigating was "bistatic" radar—a system that separated a radar's transmitter and receiver functions. In this scenario, a UAV hosting a receiver obtains intruder track information from a transmitter located on the ground, on a ship, or pretty much "anywhere," said Lacher. This was among the range of possibilities that existed in 2014. "I find it hard to believe that any single sensor will be used to cover the gamut of safe self-separation and collision avoidance; it's going to be a mosaic of sensors," Lacher surmised. "It's not going to be a 'big bang' solution. I think the solution is going to happen in stages. We're likely to see public-use airborne sense-and-avoid before we see a civil certified [system]. But a civil certified detect-and-avoid system that operates in all classes of airspace under all weather conditions? I think we're probably a decade away."

Ground-Based Sense-and-Avoid

True to its historical significance to the industry, the military led the way in breaking the civil airspace barrier that prevented wider use of unmanned aircraft. The nation's production ramp-up after the September 11, 2001, terrorist attacks had a side effect for the Department of Defense: it created pressure to absorb unmanned aircraft back home as the major conflicts in Iraq and Afghanistan subsided. The military's large number of fielded machines drove "a strong demand for access within the National Airspace

System," the DOD's undersecretary of defense for acquisition, technology, and logistics reported to Congress in April 2012:

> This need for airspace access to test new systems, train operators and conduct continental United States-based missions has quickly exceeded the current airspace available for military operations. The situation will only be exacerbated as units return from overseas contingencies…UAS will not achieve their full potential military utility unless they can go where manned aircraft go with the same freedom of navigation, responsiveness and flexibility.[17]

When the DOD conducted unmanned aircraft operations outside of restricted, warning, or prohibited areas, it needed the FAA to grant a certificate of authorization, which was only a temporary solution. "The COA process is adequate for enabling a small number of flights but does not provide the level of airspace access necessary to accomplish the wide range of DOD UAS missions at current and projected operational tempos," the Pentagon said. Moreover, "obtaining a COA requires a significant amount of time and resources, both to complete an application and work through the FAA approval process." At the time, military departments were in possession of eighty-eight active COAs at different locations around the country, most of which provided passage to a restricted or warning area. The DOD's 2012 report listed 110 potential UAS basing locations in thirty-nine states, as well as in Guam and Puerto Rico.

The Army's planned inventory of UAS in fiscal year 2017 was 7,074 hand-launched AeroVironment RQ-11B Ravens, 408 AAI Textron RQ-7B Shadows, 152 General Atomics MQ-1C Gray Eagles, and 45 Northrop Grumman MQ-5B Hunters, according to the report. By 2015, it required 2,057 total operators, 1,173 mechanics, and 316 warrant officers, the three "military occupation specialties" the Army designated for UAVs. The service's modus operandi when flying unmanned aircraft on the home front was to shadow them with a chase plane or an Apache attack helicopter, or keep them within line of sight of a trained observer on the ground. It could not fly them at night. In twenty-five locations, it required COAs from the FAA to fly Gray Eagles, Hunters, and Shadows from their launch sites to adjacent restricted areas. The 4.2-pound Raven—part of a "backpackable" aircraft system—needed only a soldier to cast it aloft, but was subject to "DOD-FAA agreed to" notification procedures for flights over government land.

With the arrival of a certifiable airborne sense-and-avoid system still years away at best, the Army started work on a technology work-around sophisticated in its own right called ground-based sense and avoid. Explaining the system's provenance, Viva Kelley, director of the Army's Unmanned Systems Airspace Integration Concepts office, said:

> Nobody invents anything in the Army; it belongs to the Army. I've got a whole team of guys that are coming up with things that nobody's ever thought of before and then the other services jump in and use them. I will say there's been a team of folks working for the Army since 2006—a small team—that really started this concept and made it successful. It was within PM UAS [the Army's UAS project management office at Redstone Arsenal, Alabama]. I can say that much.

The Pentagon designated the Army as the lead service for GBSAA, an approach the Air Force and the Marine Corps were also advancing for their UAS training needs. The Army was assisted by MIT Lincoln Laboratory, Johns Hopkins Applied Physics Laboratory, and Syracuse Research Corporation, which developed GBSAA underlying technologies.[18]

The Army envisaged GBSAA as an alternate means of complying with the FAA's Part 91.113 aircraft right-of-way rules. In essence, the system provided safe separation of unmanned aircraft from the ground, replacing what a pilot would do in the cockpit. Consisting of sensors and displays, it used ground-based radars to track aircraft. A GBSAA installation generally required two radars that supplied aircraft position information in three dimensions—range, bearing and altitude—working in concert with a site's existing ASR-11 air traffic control radar. The LSTAR 360-degree electronic scanning radar that SRC supplied to the Army for its GBSAA system "will see a one-square-meter target for greater than 40 clicks in 3D," Kelley said. "It will do better than that." The ASR-11 was a combined primary and secondary surveillance radar—respectively, one that transmits and receives radio signals reflected off airborne targets and one that interrogates transponders on aircraft and receives their replies. The radars fed tracking information to a GBSAA processing unit, including a fusion tracker that combined data from multiple radars into a single track. The system classified airborne targets to distinguish actual threats and applied a "maneuver" algorithm to prevent any collision. Radar, processing, and display elements interconnected through a "cloud" (software middleware layer) called the data distribution service.

The system's maneuver algorithm drew from the FAA's ACAS X development, Kelley said. "A, we were able to leverage work that had already been done by probably the best experts in the field and B, we also now know that it's compatible with ACAS X, which is what they're implementing into the NextGen system," she said. "But, with that said, we have put a lot of time and effort into those ACAS algorithms…to keep them from being just focused on collision avoidance and more focused on separating unmanned systems from the rest of the traffic. We've had to do a significant amount of work to make up for the man" in the cockpit.

The Army tested a prototype GBSAA system at the General Atomics flight test facility in El Mirage, California, in April 2011, under an FAA certificate of authorization. It conducted the first flight of an unmanned aircraft in national airspace using a GBSAA system on April 27, 2011.[19] The service then deactivated the system and relocated the testing to Dugway Proving Ground in Utah, where since 2009, the Army had operated a Rapid Integration and Acceptance Center to test payloads and technologies and speed their deployment to the field. In June 2012, the UAS project management office held two weeks of demonstrations to "validate" or prove the system's functionality with actual flight tests. The demonstration covered several "vignettes" involving RQ-7 Shadow and RQ-5 Hunter UAVs, as well as simulated UAVs and intruder aircraft at Dugway. The testing also replicated the airspace over other military installations and used live and recorded air traffic data from Salt Lake City and Boston Logan airports, respectively. At the time Kelley declared the demonstration a "total success."

The "Phase 2, Block 0" iteration the Army demonstrated would serve as the baseline GBSAA system for tracking Predator-based Gray Eagles—which had an all-up weight of 3,600 pounds—flying from their base airfields through military Class D airspace to nearby test ranges for training purposes. It provided the system's ground-based operator with a display of nearby air traffic color-coded to reflect the highest-priority potential conflicts. The ground operator then relayed information to the aircraft operator. Planned software block upgrades would bring new functionality: Block 1 in 2016 would provide the ground-based operator with a recommended maneuver to communicate to the aircraft operator; one year later, Block 2 would provide recommended maneuvers directly to the aircraft operator working in the aircraft's ground control station.

GBSAA could also see small Raven and Shadow UAVs, Kelley said. "Depending on the range, there are more errors associated with range, and how far out and how large you can see things becomes classified, but definitely we can see those things," she said. The system could track flying objects "where we would need to. If we're deconflicting airspace we would need to deconflict everything that's flying in that airspace."

The Army planned to test GBSAA subsystems working in concert for several months at an MIT systems integration laboratory in Lexington, Massachusetts. The service then planned to roll out GBSAA systems at five bases between fiscal years 2014 and 2016, beginning with Fort Hood, in Killeen, Texas, and followed by Fort Riley, Kansas; Fort Stewart, Georgia; Fort Campbell, on the Kentucky-Tennessee border; and Fort Drum, in upstate New York. The Army later amended that plan "due to changes in priorities based on operational needs," replacing Fort Drum with Fort Bragg in North Carolina as the fifth funded GBSAA site. It conducted a site survey with the aim of establishing a sixth location at Fort Wainwright, near Fairbanks, Alaska. By the end of 2015, the service had erected towers and finished installing radars and other hardware at Forts Hood and Stewart.

Among sites the Air Force identified for GBSAA installations were Beale, Edwards, and Palmdale air force bases in California; Creech AFB in Nevada; Cannon and Holloman AFBs in New Mexico; Grand Forks AFB in North Dakota; Ellington Field Joint Reserve Base in Texas; and Hickham Field, in Hawaii.[20] Marine Unmanned Aerial Vehicle Squadron 2, which operated the RQ-7B Shadow, planned to use GBSAA at Marine Corps Air Station Cherry Point, North Carolina. In 2014, VMU-2 became the Marines' first squadron to operate the new RQ-21A Blackjack, based on Insitu's Integrator UAV. Kelley said:

> If we had a wish list of how we could do this, then it would be a system that's integrated with airborne sense-and-avoid and it's fully integrated into the ground control station so that the operators of the unmanned system are using sense-and-avoid just like it's part of their regular system. Ground based sense-and-avoid would be in terminal areas or high density areas where you take off and land or you need to come and fly low. And then they could use the airborne sense-and-avoid part for further out in Class A airspace or less congested airspace…Ground based sense-and-avoid is a first step, but that's the ultimate vision.

In its first "Integration of Civil Unmanned Aircraft Systems in the National Airspace System Roadmap," the FAA committed to issuing a draft advisory circular on GBSAA systems and requirements by 2015, and to approve GBSAA at one or more DOD sites using Army and Air Force-developed systems, "subject to timely application and completion of certificate of waiver or authorization or other approval process." The agency planned to grant approvals "for educational and other public applications" in the period 2016 to 2018. It would draw from prior military, public, and limited civil experience to drive the development of GBSAA "minimum aviation system performance standards," leading to approved civil operations.

Maj. Gen. William Crosby served as the Army's aviation program executive officer before retiring in January 2014, after thirty-five years of active duty. During his last meeting with reporters at an Association of the US Army symposium in Arlington, Virginia, UAVs came up repeatedly. Was he confident the Army could accommodate its returning unmanned aircraft from Afghanistan and continue training with them at home? "Initially, everybody said, 'oh no, there's no way you'll ever be able to fly,'" Crosby related. "We've got restricted airspace around most of our training areas. We've been able to adapt routes that, based on the slant range, the potential of crashes, all those sorts of things, allows us to get in out." On a cautionary note, Crosby added: "We can ill afford to get something wrong here. The FAA, God love them, is partnered in working with us. As much as a pain in the butt they are—they're paid to be a pain in the butt because they protect us. But we can't afford to screw it up because then everybody is going to say 'stop.' We need to be very sequential, very positive in demonstrating how well we can manage this or we'll lose that ability to train."

Command and Control

UAVs are tethered to their ground control stations wirelessly through C2 and mission-related or "payload" radio links. The C2 link used to control the aircraft was considered safety critical. Through this invisible pipe the pilot transmitted flight commands to the UAV; the UAV sent back telemetry data from on-board instruments with aircraft position and status information, data for the pilot's navigation displays, and target data from the aircraft's sense-and-avoid system. Separate of the C2 link, a communications loop connected the ground station, the aircraft, and air traffic controllers, providing controllers the means to issue voice or data message instructions to the UAV pilot. The payload link was the pipe devoted to the particular mission payload, rather than to controlling the aircraft. It accommodated sensor commands uplinked to the aircraft and sensor signals downlinked to the mission system operator.

Carving out the needed radio frequency spectrum to host those links, and especially safeguarding the C2 link from interference or manipulation, was prerequisite to introducing large numbers of unmanned aircraft into the national airspace system. As those operations increase, "so will the needs of UAS CNPC (control and non-payload communications) links for radio spectrum," MITRE researchers advised in a 2013 paper. "...Future CNPC bandwidth requirements will be determined by the message transfer requirements of individual (unmanned aircraft), and by the numbers, densities and distributions of future UAS deployments."[21]

Radio frequency spectrum was "harmonized" (coordinated) on an international basis through the International Telecommunications Union, a United Nations agency based in Geneva, Switzerland. Coordinating spectrum assignments among nations protected against radio frequency interference, ensured adequate bandwidth was available for certain applications, and promoted their operation across international borders. Every three to four years, the ITU held a World Radiocommunication Conference to review or revise its regulations governing the use of radio frequency spectrum. At the WRC in February 2012, negotiators reserved aviation safety protected spectrum in the L band (960-1164 MHz) and the C band (5030-5091 MHz) for unmanned aircraft C2 non-payload links. Those bands supported line-of-sight, ground-based links; identifying spectrum for beyond line of sight, satellite communications was on the agenda for the WRC in 2015.

The L band also supported distance measuring equipment (DME) and shipborne tactical air navigation radionavigation systems, but there was capacity in the 960-977 MHz subband for C2 land links, MITRE researchers found. "That's really good quality spectrum; it has a lot of good propagation characteristics. The radio line of sight can go out to about 200 miles," said the FAA's UAS Integration Office manager Jim Williams. The C band allocation was originally intended for the microwave landing system, which the FAA once intended to replace the instrument landing system for conducting precision approaches to runways—until it was overtaken by the GPS Wide Area Augmentation System. "The C band has much lower propagation. It's probably only good to 30 to 50 miles depending on whether you have a directional antenna or not," Williams said. "Between those two bands, we figure we can accommodate most of the known needs" for unmanned C2 links.

Under a $2.45 million cooperative agreement NASA's Glenn Research Center awarded to Rockwell Collins in 2011, the avionics manufacturer developed a UAV control and non-payload communications "data link" that connected pilots in ground stations with one or multiple unmanned aircraft. In turn, UAVs could communicate with more than one ground station, relaying from one to another much like cell phones. Rockwell Collins' prototype radios talked to each other in the L and C bands designated for unmanned aircraft. The unique radio "waveforms" (signals) the project created would be made available as a "public resource" the FAA could use in developing a common C2 standard that industry could build systems around. The partners conducted flight tests in Ohio and Cedar Rapids, Iowa, in 2013–2014, using NASA Glenn's S-3B Viking twin-engine jet and other aircraft as surrogate UAVs to test the system. The testing was to continue through 2016.

UAV sensor payload operators would have to find spectrum elsewhere. Most such operators used unlicensed bands, such as 2.4 GHz—typically used for wireless gadgets including routers, cordless telephones, and garage door openers—and the less congested 5.8 GHz band, both of which fell within the purview of the Federal Communications Commission. "The payload links are not (for) safety; command and control links are," Williams said. "I think the real spectrum crisis for the industry is there's not enough spectrum available for payload communications. We're not going to use safety-protected spectrum for that, period, end of story."

C2 link reliability was another challenge the industry faced. In theory as well as practice, a UAV that somehow became untethered from its C2 link followed a pre-programmed contingency route until the link was reestablished or the operator safely ended the flight. "The intent of any lost link procedure is to ensure airborne operations remain predictable," according to the FAA, which required that operators provide contingency procedures as part of their COA.[22] Nevertheless, there were several well-publicized incidents in which the "lost link" scenario did not play out as planned. In August 2010, an MQ-8B Fire Scout operated from Patuxent River Naval Air Station, Maryland, along the Chesapeake Bay, flew into restricted airspace near Washington, DC, an incident the Navy blamed on a software glitch. Rather than returning to base as programmed, the unmanned helicopter strayed twenty-three miles on a north/northwest trajectory toward the nation's capital. The operators managed to regain control of the aircraft by shifting to another ground station, the Navy said at the time.[23] A NASA Aviation Safety Reporting System sampling of forty-four UAV "events" between 2003 and 2013 contained reports of human-induced programming errors and unexpected flight plan deviations submitted by operators, as well as TCAS alerts and surprise flybys noted by pilots. The synopsis of one such event in 2012 stated: "While attempting to climb above weather, the pilot of a UAV lost control of his aircraft at FL210 [21,000 feet] and lost altitude to 16,500 MSL [mean sea level] before regaining control and restarting the climb." Another from a 2011 event stated: "MQ-9 pilot reports ground station computer lock-up which results in the aircraft flying outside its designated boundary. When the terminal is turned off the aircraft returns to its emergency lost link mission. Another ground station is used for landing." NASA scrubbed identifying details from the reports, which were submitted voluntarily.[24]

Lost link incidents made for unpredictable airspace operations, increased the risk of airborne collisions, and heightened controller workload, the FAA said. "Lost or degraded link events—for either control or voice communications—currently occur too frequently for (national airspace) integration," the agency advised in its UAS "Concept of Operations" version 2.0, dated September 28, 2012. Introducing UAVs into the airspace system "is precluded by the unknown reliability of the communications links and the lack of uniform procedures to respond to link anomalies."[25]

Willful interference of UAVs in flight was another issue. In a September 2012 report to Congress, the Government Accountability Office fretted that GPS signals unmanned aircraft used for guidance could be easily jammed with low-cost devices, affecting their control. A countermeasure would be to equip the aircraft with a second or backup inertial navigation system—an option for large air vehicles, but not for smaller ones with tighter tolerances for size, weight, and power. A small UAV might beat the threat of GPS jamming by encrypting its communications, but the cost and weight penalties might also make this infeasible.

GPS "spoofing," the act of transmitting a false or counterfeit GPS signal toward an aircraft's receiver to overtake its navigation, was an emerging issue, the GAO said. The agency cited a demonstration University of Texas at Austin researchers conducted that year at White Sands Missile Range, New Mexico. Led by engineering assistant professor Todd Humphreys, the team hijacked an Adaptive Flight Hornet Mini unmanned helicopter from about 0.62 miles (one kilometer) distance "and induced it to plummet toward the desert floor." In a press release, the university claimed credit for "the first known unequivocal demonstration that commandeering a UAV via GPS spoofing is technically feasible." Said Humphreys: "We're raising the flag early-on in this process so there is ample opportunity to improve the security of civilian drones from these attacks, as the government is committed to doing."[26] Adaptive Flight issued its own press release, saying it was proud to participate in the demonstration, but also noting that the patent pending "GPS-denied" mode of the Hornet Mini's flight control system "was turned off, as the intent of the test was to have a UAV configured to rely on GPS."[27]

As Humphreys suggested, the FAA's UAS Integration Office was committed to safeguarding the C2 link between UAVs and their ground control stations from unwanted intrusions. The agency hired Leidos, of Reston, Virginia, to provide the information security component of a C2 data link standard then being developed by an RTCA special committee. Leidos—a play on the word "kaleidoscope"—was formerly the national security, engineering, and health businesses of federal contractor SAIC before the two entities separated in August 2012. "These are guys who support the US Navy in their cryptology world, so we're essentially funding the expertise that the military has built," Williams said. "That particular piece is designed to protect against somebody taking over the link, hacking into and taking it over, which a lot of people claim that they can do," he added. "The idea is that we want to prevent that from happening to a high enough level of assurance that it maintains the safety of the system."

In other areas of UAS airspace integration the FAA lagged—including, importantly, training its controller workforce to manage the coming wave of unmanned aircraft. As of mid-2014, the agency had provided an hour-long computer-based UAS course but little other formal training, said Chris Stephenson, a former Navy and civilian controller who served as UAS subject matter expert with the National Air Traffic Controllers Association; Stephenson was the only UAS expert at the union's Washington, DC, headquarters at the time. Nevertheless, controllers at several of the FAA's air route traffic control centers tasked with managing high-altitude traffic had safely separated military Global Hawks and Customs and Border Protection Predators for years, "and they're kind of each doing it their own way as far as training," he said.

Those public operators had some hiccups along the way. An MQ-9 that a General Atomics pilot flew for the CBP crashed into terrain northwest of Nogales International Airport in Arizona on April 25, 2006, an accident the National Transportation Safety Board attributed to crew errors. In June 2012, a Navy RQ-4 Global Hawk demonstrator crashed in a marsh near Salisbury, Maryland. In January 2014, the CBP temporarily grounded its Predator fleet after operators ditched a malfunctioning Guardian in the Pacific Ocean near San Diego. According to Stephenson, earlier-generation Global Hawks would start an immediate climb as part of their C2 lost link procedure, which posed a problem for controllers trying to maintain vertical separation among nearby aircraft. The RQ-4 Block 40 Global Hawk came with an altitude hold capability that resolved the issue, and the Air Force upgraded its older aircraft so that special procedures for controlling them became unnecessary, he said. "Right now, fortunately, where all of these things are flying there's not a whole lot of other traffic. Basically the rules are such that you work it just like any other aircraft," he explained. "If you actually talk to the controllers, they'll tell you that if I have a manned aircraft and an unmanned aircraft, one of them has got to move. The manned aircraft is going to move because they have a lot more confidence in that. But that's not in writing anywhere; that's just the way they're doing it right now."

A June 2014 audit report the Department of Transportation Inspector General delivered to Congress affirmed Stephenson's account of the training situation. The report stated:

> Currently, although FAA has authorized some UAS to operate in the national airspace system at select locations, such as along the nation's borders, the agency has not developed the procedures, training and tools for controllers to effectively manage UAS in the same airspace as other aircraft. [C]ontrollers at one air traffic control facility handling large UAS operations told us that they always moved manned aircraft away from UAS because they were not aware of the specifics of individual UAS missions and performance characteristics, such as differing airspeed capabilities and rates of climb.[28]

The IG noted a simulation the FAA's Hughes Technical Center conducted in 2011, that mixed multiple, simultaneous operations of unmanned aircraft with manned aircraft in Class D airspace—generally airspace from the surface to 2,500 feet around airports with control towers. According to that study:

> Dissimilar flight characteristics and the inability for UAS pilots in command to visually acquire and follow other aircraft in the traffic pattern were observed as a main causal factor for delays, loss of situation awareness, extended patterns, increased air traffic controller workload, and a general reduction in overall safety…The presence of UAS contributed to increased ATC workload especially with more than one UAS present. [The] limited scope operational assessment of multiple simultaneous UAS operations in Class D airspace suggests the proposed operations were, in general, not feasible as simulated."[29]

Stephenson was optimistic the FAA and its controllers would master the new unmanned aircraft entrants to the airspace. "We had our tough times as far as labor relations go between NATCA and the FAA but we got past it and we're in a very collaborative mode now," he said. "We've got a good administrator and a good [union] president here that are on good terms and so we're involved."

Developing Standards

One year after President Obama signed the FAA Modernization and Reform Act of 2012, the UAS community had misgivings. In the watershed legislation, Congress directed the FAA to open the nation's airspace to unmanned aircraft by September 2015, but progress toward that end seemed to be at a standstill. The FAA planned to honor the congressional mandate by introducing UAVs incrementally, beginning with small air vehicles weighing less than fifty-five pounds. But the agency had still not issued its small UAS draft rulemaking, which was apparently mired in federal bureaucracy. "Just when we thought we were making some progress, we got blind-sided by the privacy issue," complained UNITE president Carl Mikeman, alluding to the storm that arose over the surveillance capabilities of "drones." Mikeman was speaking to a friendly audience at the time—an AUVSI Oklahoma chapter event in Norman, Oklahoma, in March 2013. Sprinkled about the crowd were UAV manufacturers, college professors, state government officials, and a few trade journalists. The state's Department of Commerce had flown in the journalists to help spread the word about Oklahoma's ultimately unsuccessful bid to host an FAA-designated UAS test range.

Introducing small UAS was one thing; gaining wider access to the airspace for large unmanned aircraft was something else altogether, said Mikeman, whose employer, Global Hawk manufacturer Northrop Grumman, had an interest in making that happen. There were technology barriers to overcome when flying an aircraft beyond an observer's line of sight, chief among them the need to prevent collisions with other aircraft. An industry and government group assembled by standards organization RTCA—Special Committee 203—had been working on overall standards and certification criteria for UAS sense-and-avoid and C2 systems since October 2004, but "progress is painfully slow," Mikeman said. In fact, he said, the community had just learned that RTCA was disbanding SC-203 and forming a new group focused more narrowly on equipment requirements for sense-and-avoid and C2.

Founded in 1935 as the Radio Technical Commission for Aeronautics and known in recent times by its acronym, RTCA was a nonprofit organization that functioned as a federal advisory committee to the FAA. It was in essence a meeting of the minds that operated in a close orbit around the agency. In response to taskings from the FAA on technical problems the agency needed resolved, RTCA formed committees of industry and government experts who met on a volunteer basis and made recommendations based on a consensus of their members. Their findings served in developing minimum operational performance standards (MOPs) and minimum aviation system performance standards (MASPS) the FAA incorporated in its regulatory and advisory documents released to industry. The standards also guided designers and manufacturers in building and certifying equipment to the FAA's satisfaction.

RTCA convened SC-203 to develop MASPS—system level standards—for unmanned aircraft and separately for their required C2 and sense-and-avoid systems; eight years later it was still at work. The enormity of the task the committee faced was evident during a plenary meeting Northrop Grumman hosted in San Diego in March 2008. James Sizemore, then the FAA's designated federal official on SC-203, "stated that the FAA understands and hears the frustration that the UAS community is having, but that setting UAS standards is a huge undertaking requiring much work," according to the meeting minutes. Kenneth Geiselhart, representing Lockheed Martin as one of two industry co-chairmen, "walked through the various options that the leadership team has looked at to shorten the process of developing the MASPS. The leadership team believes that the time could be shortened up to two years by following Option 4." Geiselhart delivered a presentation on resources the committee needed to accomplish its tasks. "He briefed that the staffing needs and requirements slide in his presentation [are] over and above the current staffing levels of SC-203."[30]

By 2012, the FAA had consolidated its unmanned aircraft activities within the UAS Integration Office, which then moved within RTCA to expedite the standards-setting process. Jim Williams was part of an "ad hoc" group the RTCA's program management committee formed that December to review SC-203's founding "terms of reference," akin to its charter. RTCA President Margaret Jenny briefed the situation to an SC-203 plenary meeting in February 2013. According to the minutes, Jenny explained the committee's focus on developing MASPS was "complex work based on the scope of the existing Terms of Reference. Ms. Jenny went on to say that there are some stakeholders, specifically the Department of Defense, who are concerned that product development is not happening fast enough."[31]

The die was cast for SC-203. "One of the problems, we all finally determined, was that we had given SC-203 an impossible task, which is to define standards for all kinds of vehicles and all kinds of missions. It was just intractable; it was just too big to get their arms around," Jenny later told *Aviation International News*.[32] "Here's the deal on 203, and I probably said this at every meeting," related Ted Wierzbanowski, who served on the committee. "You guys are trying to do too much. You've expanded the scope. You're trying to solve world hunger…With the direction they were getting, there was no way they could be successful." Williams saw the problem and moved to resolve it, he said.

In March 2013, RTCA's program management committee resolved to create a successor special committee—SC-228—comprising both sense-and-avoid and C2 working groups. These would focus on a subset of UAS and a subset of their missions, generally larger unmanned aircraft transitioning through lower altitudes to Class A airspace above 18,000 feet, or to special use airspace, Jenny said. The "PMC" (management committee) believed the sharper focus would hasten the development of MOPS—standards used in designing equipment—and facilitate the Pentagon's need to move unmanned aircraft from their bases to training ranges and the FAA's requirement from Congress to introduce UAVs into the national airspace system by September 2015. "PMC discussion emphasized that the re-focus of RTCA UAS committee activities was a difficult decision and that the PMC is very grateful to SC-203 and the committee's leadership for the progress achieved to date," meeting minutes stated. "That the SC-203 Terms Of Reference was too broad was a PMC fault and not a committee shortcoming... and the PMC decision not to retain any of the current leadership was to avoid sending an unintended message."[33]

In the organization's parlance, RTCA "sunsetted" SC-203 in June 2013. John S. "Johnny" Walker, the long serving SC-203 co-chairman, former FAA executive, and unmanned aircraft proponent, said the committee's work had not been for naught, but served as guidance for the successor committee and for its European counterpart—Eurocae Work Group 73—as well as for ICAO's unmanned aircraft study group. "Although we are morphing into a new committee focused on MOPS, the foundational work is in place," said Walker, who sounded disappointed during a telephone call. "It's not like gee whiz, what have you done for seven years, and (then) throw it out."[34] Rose Mooney, who served as co-chairman with Walker, recommended SC-203 be retained to continue working alongside the new committee. "I thought what they were doing with 228 was a good exercise, but we thought it should be done in parallel," she said. "They've narrowed the field to implement a couple of technologies—the Air Force sense and avoid (system) and the Rockwell Collins radio and they're doing it for Class A airspace access. But all of the risk and hazard analysis that SC-203 was working toward will still have to be done... Yes it was taking way too long to do what we needed to do and it was a big scope, but we were well on the way."

With aeronautical communications consultant George Ligler and Insitu vice president of government relations Paul McDuffee named co-chairmen, SC-228 held its first plenary gathering at RTCA's downtown Washington, DC, offices on July 30, 2013. The manifest listed 169 names of people who attended the public meeting. The committee's marching orders were to produce final MOPS documents by July 2016.

Paul McDuffee, a former Embry Riddle Aeronautical University professor and chief flight instructor, was Insitu's vice president of government relations and strategy. He was the Boeing subsidiary's principal liaison with the FAA on matters related to regulations, and he co-chaired RTCA Special Committee 228 on unmanned aircraft. *Author photo.*

A year later, SC-228 was itself challenged. "That's the $64,000 question," McDuffee said when asked if the committee would deliver the UAS performance standards on schedule. "We are doing an agonizing appraisal now of where we are and where we're not, and where we need to be." The committee's plan was to deliver draft standards for detect-and-avoid and C2 by the following year, but it was experiencing "spotty and inconsistent participation" by the FAA, whose full cooperation SC-228 needed to fulfill its task. Without that involvement, the options were to slip the deadlines or to potentially suspend the committee. "We've got 200 people on Special Committee 228, all volunteers, spending a lot of their good time and treasure to come up with a quality product, and the last thing we want to do is present something that's useless," said McDuffee, who joined Insitu after three decades as a professor and chief flight instructor at Embry Riddle Aeronautical University. "We're going to do everything we can to meet those deadlines—that's an absolute commitment. But there will come a point in time where without adequate and consistent support from the rulemakers we may end up delaying things to the point where we can't deliver. That's a problem."

By the end of 2014, McDuffee had amended his outlook. "I can report that we are making tremendous progress. We are roughly a year away from actually having a first draft of the performance requirements which will ultimately be verified and validated by NASA and others to make sure we're on the right track," he told an Air Line Pilots Association conference in December. But he tempered his remarks, cautioning that any technology solutions would have to be affordable and relevant across different operators and UAVs. "We don't believe there's going to be a one-size-fits-all solution to any of this," he said. Keeping to its schedule, the committee completed draft standards for command and control and collision avoidance in September 2015.

CHAPTER 9

THE STATES WEIGH IN

The fiscal 2012 National Defense Authorization Act that President Obama signed into law in December 2011 and the FAA Modernization and Reform Act of 2012 he signed several weeks later required the FAA to select six national sites or "ranges" to test unmanned aircraft. Congress provided no federal money for the program, which was to last five years. Using the same language, the bills instructed the FAA to designate the airspace for the ranges, develop the certification standards and air traffic requirements needed to use them, coordinate the program with NASA and the Department of Defense, address civil and public unmanned systems, ensure the program aligned with NextGen, and verify the safety of the machines before releasing them to the sky. Both bills directed the FAA to establish the test site program within six months of their enactment—roughly by the summer of 2012—and to activate the first range six months after that, or in early 2013. The agency was to consider "geographic and climatic diversity" and "the location of ground infrastructure and research needs" when choosing which states would host the ranges.

The sites could help the FAA address sense-and-avoid, command and control, and other challenges preventing wider use of unmanned aircraft. In timely fashion, in March 2012, the agency published a request for comment on the site selection process in the *Federal Register*. It hosted two public webcasts that April. "Ultimately," as a result of that outreach effort, "the FAA elected to implement the test-site program by selecting six non-federal, public entities to operate the test sites with appropriate FAA oversight," Administrator Michael Huerta explained in a written order.[1] But a full year transpired from the date President Obama signed the FAA reauthorization legislation until the agency issued a "screening information request"—a request for proposals—to operate the sites in February 2013. The official explanation was that controversy over privacy slowed its progress. "FAA officials told us they…had expected to issue a request for proposals in July 2012," the Government Accountability Office reported that fall. "However, because of privacy concerns regarding the collection and use of UAS-acquired data expressed by commenters, the internal review process was delayed."[2]

Half the country submitted proposals for the FAA's test site solicitation. The agency reported receiving twenty-five submissions from public entities and universities in twenty-four states. While no federal money would be forthcoming, test site aspirants viewed the FAA's endorsement leading to an economic cluster not just around testing unmanned aircraft, but designing, building, equipping, and maintaining them. The aircraft were almost secondary—they needed control stations, communications equipment, ground radars, simulation and training devices, and software to shape and analyze the data they collected. "We always looked at this as the science necessary to get to the commercial application," said Lawrence Brinker, executive director of a group of aerospace companies and academic institutions in New York and Massachusetts called the Northeast UAS Airspace Integration Research Alliance (NUAIR). The two states were represented in Washington, DC, by the same lobbying firm—The Roosevelt Group—which recommended they combine their proposal to the FAA. "Obviously, where we can develop companies in Massachusetts and New York in this industry segment—that's what we want to do," Brinker said.

Congressional delegations and state and local governments lined up behind the proposals. Nevada Gov. Brian Sandoval convened the Nevada Autonomous Systems Policy Planning Panel, a blue-ribbon committee of industry, academic, and government executives, to plan that state's application. The legislature appropriated $4 million for an "Unmanned Aerial Vehicle program" which could only be allocated "upon Nevada's successful designation" as an FAA test site. North Dakota Gov. Jack Dalrymple assembled the Northern Plains Unmanned Systems Authority and appointed Robert Becklund, a former Air National Guard colonel, to run it. The North Dakota Legislature approved $5 million to develop a test site, with $4 million contingent on winning the FAA's nod. There was "fierce competition" for the FAA endorsement, said Stephen McKeever, Oklahoma secretary of science and technology. Oklahoma had a compelling case to make: it had its own governor's UAS advisory council and it already hosted the Department of Homeland Security's Robotic Aircraft for Public Safety demonstration range. Oklahoma State University laid claim to being the nation's first institution of higher education to offer master's and doctoral degrees with UAS options. The University of Oklahoma was recognized as a leader in radar and weather research using unmanned aircraft. The state had access to 200 square miles of restricted airspace at Fort Sill to safely test unmanned aircraft, as well as other locations. It mounted an impressive public relations campaign. In March 2013, Oklahoma industry and government officials serenaded reporters at a reception the state's commerce department hosted on the 49th floor of the nearly new Devon Tower in downtown Oklahoma City. Gazing down at the meandering Oklahoma River below, it seemed the state was a shoe-in for the FAA's coveted stamp.

The FAA announced its test site selections on December 30, 2013, keeping a promise it had made repeatedly to name the sites by the end of the year. The lead organizations

North Dakota was one of the states the FAA chose to host an unmanned aircraft test site after a hard-fought competition. Robert Becklund, executive director of the Northern Plains site, kept this commemorative license plate in his office at the University of North Dakota at Grand Forks. *Author photo.*

representing the winning sites were the University of Alaska-Fairbanks, the state of Nevada, Griffiss International Airport in Rome, NY, the North Dakota Department of Commerce, Texas A&M University-Corpus Christi, and Virginia Polytechnic Institute and State University. Entities from other states would also participate in the test sites, which typically combined multiple locations. Alaska's Pan-Pacific UAS Test Range included sites in Oregon and Hawaii. Virginia Tech's Mid-Atlantic Aviation Partnership included the University of Maryland and Rutgers University in New Jersey. Griffiss Airport was the lead applicant for the NUAIR alliance of New York and Massachusetts. The proponents of the winning sites rejoiced. "Being selected as one of six sites for UAV development in the country is a historic moment for #NV. This selection is a major win for #NV!" Sandoval tweeted following the announcement. Home to Creech and Nellis air force bases, Nevada considered itself the birthplace of unmanned aviation. "We're building and growing an industry cluster that already exists here," said Tom Wilczek, aerospace and defense industry specialist with the Governor's Office of Economic Development. "It was very important for us to achieve the FAA designation for a couple of reasons. No. 1, we felt it validated our place in the industry. Also, to us, it meant that the FAA really believes Nevada can stand up and help them in meeting their congressional mandates."

In a promotional video Ron George, Texas A&M senior research development officer, equated his university's role to the Wright brothers' historic first airplane flight. The opportunity to introduce unmanned aircraft in the nation's airspace "is absolutely a Kitty Hawk moment for the history of this institution," he said. "We're talking about a multibillion-dollar economic impact here." As lead applicant of the Lone Star UAS Center, Texas A&M held a "citywide celebration" at the Brewster Street Icehouse in downtown Corpus Christi. The sun-bleached city on the Gulf of Mexico provided financial support to the test site campaign. "We're indebted to them, and they're proud," George told the author. "We looked upon this technological area as a niche to become the best in the country. We're committed to doing that."

Mostly university oriented, the test sites were based in states or regions that were generally known as conservative, each with a strong heritage of military service. "Nevada in general looks at this industry as our birthright," explained Wilczek. "I think we have a higher proportion of residents that have supported the military and this technology. In terms of other states, there is a much higher proportion of individuals here who understand these platforms." The sites were closely linked with the military establishment or retired military officers in their respective states.

Robert Becklund commanded the North Dakota Air National Guard's 119th Fighter Wing at Fargo from 2004 to 2009, a time of transition that saw the wing retire its F-16A Fighting Falcons as a result of the Pentagon's 2005 Base Realignment and Closure (BRAC) recommendations. In its place, the wing accepted responsibility for the unmanned MQ-1 Predator. "I was the guy that got us safely out of the F-16 and transitioned into the Predator," he recalled. "We made a conscious decision to set ourselves up for the BRAC in 2005 to get the Predator. It was like a proactive engagement in our future." Becklund then worked at the Pentagon for three years on the Air Force's remotely piloted aircraft task force. Maj. Gen. Robert Knauff, a fighter pilot who commanded the New York Air National Guard from 2005 to 2009, was NUAIR's founding CEO. Tragically, he was killed in a glider accident in Arizona in May 2014, before the test range started operating. Lawrence Brinker was a former Air Force pilot and aviation lawyer who for a time advised the commander of the US Special Operations Command. Rosanne "Ro" Bailey, director of Alaska's Pan-Pacific UAS Test Range Complex, was a retired Air Force brigadier general who ultimately served as commander of the Cheyenne Mountain Operations Center in Colorado.

Some of the sites were analogous with the nation's top oil-producing states. Texas and North Dakota were the top two state producers of crude oil in 2013, respectively, with Alaska and California tied for third, according to the US Energy Information Administration. The federal offshore region of the Gulf of Mexico was responsible for 17% of overall production.[3] The Bakken oil fields of western North Dakota were experiencing a historic boom, and oil companies were poised to tap the Arctic Ocean to supplement declining production from Alaska's North Slope. Unmanned aircraft figured to play a supporting role in monitoring oil pipeline and other infrastructure on land and marine mammals at sea. Addressing AUVSI's Unmanned Systems Conference in May 2014, Mead Treadwell, then Alaska's lieutenant governor, noted the FAA had established permanent operational areas for unmanned aircraft in the southern and northern Arctic and the Bering Strait. "We're very excited about this," he said. "It will help these vehicles be a major part of what's happening with oil and gas exploration in our state." Shell had invested billions of dollars exploring the Berger Prospect in the Chukchi Sea, Treadwell added. "Frankly, we want to make sure that we don't have bad interactions with wildlife. Unmanned aerial vehicles are a tremendous opportunity to do that more thoroughly and more safely and more economically."

Naming the test sites was only the beginning of another process; the FAA's rolling interpretation of congressional intent called for "standing up" the first site within six months of the selections. The sites had to first enter into "other transaction agreements" with the FAA to formalize their relationship with the agency to conduct research. The OTAs established binding terms and conditions for operating the test sites and required, among other supporting material, a privacy policy. The issue of protecting privacy rights from overhead surveillance—all-consuming at the national level—was not a significant obstacle for the test sites. The Lone Star UAS Center had state legislation behind its privacy policy. Under the Texas Privacy Act that Gov. Rick Perry signed into law on June 14, 2013, it was illegal to use an unmanned aircraft to capture an image of an individual or privately owned property "with the intent to conduct surveillance," and also to possess or distribute the image. However, the law specified nineteen legitimate uses of drones, including for "scholarly research," for police purposes with a warrant or in certain situations, and specifically, for operations within an FAA-designated test range. The center had "participated in the development of that law by providing information to legislators regarding proper uses of UAS that ought to be exempt from the law's provisions."[4] In North Dakota, a nineteen-member UAS Research Compliance Committee consisting of University of North Dakota appointees and representatives of the Grand Forks government, emergency response agencies, and the greater community reviewed and approved all research projects using unmanned aircraft. Nevada addressed privacy through a "combination of good policy married with the technology," Wilczek said. "We actually embraced that requirement and are very much looking forward to assisting the FAA," he added.

Privacy policies notwithstanding, partnerships the states cobbled together for their test site applications had to quickly become working entities. "Each test site had a variety of issues," said Rose Mooney, who Virginia Tech hired in April 2014 to lead

Virginia Tech appointed Rose Mooney to lead the Mid-Atlantic Aviation Partnership in April 2014. Mooney previously headed the airspace integration efforts of AAI Textron, led the UNITE partnership of major UAS manufacturers, and co-chaired RTCA Special Committee 203 before it was disbanded. *Author photo.*

the Mid-Atlantic Aviation Partnership. "One is we had to figure out how to do business. We're embedded in universities, many of us. We had to get staffing in place, we had to get funding in place. There were a number of things…in regards to how do you operate a small business in the middle of a university?" Mooney arrived with industry acumen and knowledge of the sausage-making process in Washington, DC. She had served as president of the UNITE partnership of major UAS manufacturers and co-chaired the ill-fated RTCA Special Committee 203. Becklund, a former fighter pilot, set up shop at the University of North Dakota-Grand Forks, amid the elevated walkways affectionately called "gerbil tubes" that shield UND students from North Dakota's bitter winters. "I live here rent-free, if you will," he said. "The Dean gives me a little office and it's fine. I don't need to go spend my appropriated money on other places."

Each site would have to go through the COA process before gaining the FAA's clearance to fly a specific unmanned aircraft. In that respect, UND had a jump on some of the other sites. The university's John D. Odegard School of Aerospace Sciences had helped secure a COA for the Grand Forks Sheriff's Department to fly the Draganflyer multirotor helicopter and was certified by the manufacturer as a training provider. At the time, cross-state rival North Dakota State University, which was a partner in the Northern Plains UAS test site, sought to employ an unmanned aircraft for conducting research in "precision agriculture," the use of aerial and satellite imagery to more effectively manage crops; the catch was that it planned to use a flying wing model, the Trimble UX5. In an example of the kind of improvisational thinking that was a hallmark

The University of North Dakota already had an FAA certificate of authorization for the Draganflyer multi-rotor helicopter that would help it start test site operations. North Dakota State University would manage the first approved research project, but NDSU wanted the Trimble UX5 flying wing, not the Draganflyer. *Author photo*.

of introducing UAVs into the nation's airspace, the test site leadership devised a workaround. "We thought: Let's talk NDSU into using the Draganflyer to begin with, so we can get their experiment going and meet the FAA's timeline for flying," Becklund said. "The Draganflyer is not their preferred alternative for aircraft; they want to use the Trimble UX5 airplane. But we thought, and probably rightly so, that if we went for the Trimble right away the COA would take too long and we would risk not being able to meet the FAA's timeline to fly. It was the easiest way to help the FAA meet their mandate and I think they're very grateful for that kind of thinking."

With North Dakota's congressional delegation and Gov. Jack Dalrymple beaming in the background, Huerta—in Grand Forks for the occasion—announced the award of the first test site COA to the Northern Plains site on April 21, 2014, ostensibly beating the congressional requirement to stand up a site within six months by more than two months. Two weeks later in Anchorage, Huerta bestowed the second COA on the University of Alaska, which that same day conducted the first test site mission by launching an Aeryon Scout quadcopter to observe musk oxen at the university's Large Animal Research Station in Fairbanks. The FAA then awarded COAs to Nevada on June 9, Texas A&M on June 20, Griffiss International Airport on August 7, and Virginia Tech on August 13.

By the fall of 2014, the enthusiasm that accompanied the original test site declarations gave way to insecurity. What actually would be their role in introducing unmanned aircraft into the airspace system? When the FAA made its selections the previous

December, it said each test site would address various research areas: Alaska's Pan-Pacific range would develop UAS air vehicle and safety standards; Nevada would focus on air traffic control procedures for unmanned aircraft, as well as operator standards and certification requirements; NUAIR was to develop UAS test, evaluation, and validation processes, and also study sense-and-avoid technology; the Northern Plains site would develop "airworthiness essential" data for unmanned machines, validate "high-reliability" link technology, and conduct human factors research; Texas A&M would develop system safety requirements for both air vehicles and operations; and Virginia Tech would conduct UAS failure mode testing and "identify and evaluate operational and technical risks areas."

While the agency communicated broad research objectives, test site directors complained it provided little other guidance to help them further their missions. Researchers were studying applications of unmanned aircraft, but not the characteristics of the aircraft themselves—the raison d'être of the test sites. At a briefing in Grand Forks, Becklund said North Dakota State University was advancing its knowledge of precision agriculture using unmanned aircraft, but the Northern Plains test site awaited direction from Washington on UAS-specific research objectives. "The thing I would like to see is the FAA would define the areas of research that they would like the test sites to focus on so that it's deconflicted and avoids redundancy," he said. "The challenge is there's no funding for this. They're very sensitive about (enforcing) a mandate upon these test sites that's not funded, so they're walking a fine line themselves. From a national point of view, that needs to be organized."

Virginia Tech's Mooney said she preferred to not criticize the FAA in public, but she acknowledged she was frustrated with the lack of guidance. "We're gathering data and we're gathering information but they have not given us a clear path on how they plan on using it," she said. By late in the year, the FAA had allocated a person at its William J. Hughes Technical Center in New Jersey to plan how the agency would make use of the test site data, "but they have been pretty nebulous on that," Mooney added.

Robert Becklund, a former wing commander in the North Dakota Air National Guard, headed the Northern Plains UAS test site from the University of Dakota at Grand Forks. Months after it started conducting test flights, Becklund warned that the site was not getting needed direction from the FAA. *Author photo*.

The FAA also floated the idea of using the test sites for something practical. The agency considered delegating authority to the sites to issue special airworthiness certificates in the experimental category—the "tickets" commercial entities needed to test, demonstrate, or train on their unmanned machines. The sites would employ "designated airworthiness representatives" for that purpose, helping the FAA process applications. Becklund was enthusiastic about the possibility. In an interview with *Aviation International News*, he said his understanding from the FAA was the test sites would become the main dispensaries of experimental certificates. "This is a place where people can come and fly and get hours on their airplanes under the scrutiny and approval of the FAA and have those hours and procedures count toward an ultimate certification either in the 'experimental' or 'restricted' or other category to be developed. Eventually, they can get their airplane certified for commercial use—that's what everybody wants," he said.

Following publication of the article, the FAA asked through an intermediary that a clarification be added. "The option is being explored to allow/create a designated airworthiness representative to assist the FAA with special certification of UASs in the experimental category," the agency stated. "The FAA will continue to process special certifications in the experimental category in accordance with its current policies and procedures. The FAA has no intention of limiting special certification for any aircraft solely to the six designated UAS test sites."[5]

Earlier in her career, Mooney had shepherded the RQ-7 Shadow through the FAA's experimental certification process as the director of engineering support for AAI Corporation. While she allowed that having designated airworthiness representatives at the test sites might benefit the sites, she was less sure of the value proposition for companies seeking to certify their unmanned models. An experimental ticket allowed a company to fly its machine for research and development, crew training, or sales demonstrations—but not for commercial gain. "I went through the experimental certificate process in 2007 with the Shadow. It was very time consuming, it was very expensive and it really yielded not much," Mooney said.

Moreover, the six test sites Congress ordered created in the 2012 FAA reauthorization legislation were not the exclusive centers of activity for testing unmanned aircraft in the United States; there were other civilian sites available. New Mexico State University at Las Cruces had operated one since 2008. Oklahoma, disappointed in its bid for an FAA-designated range, co-founded the International Consortium of Aeronautical Test Sites with organizations in France, the United Kingdom, and Canada. It retained the Department of Homeland Security's RAPS program. Space Florida, that state's aerospace economic development agency, also missed the FAA's cut, but vowed to stay involved. "Much of the country's UAS industrial growth will occur outside the states selected today," Frank DiBello, the agency's president, said in a press release after the test site selections. "Florida is still in the game." The Ohio/Indiana UAS Center, a joint venture of those two states based in Springfield, Ohio, obtained authorizations from the FAA to fly UAVs for precision agriculture at Ohio State University's Molly Caren Agriculture Center. It had a partnership with Springfield-Beckley Municipal Airport, which hosted hangar space for unmanned aircraft.

Another challenge arose that threatened to undercut the mission of the six test sites. The exemptions the FAA granted under Section 333 of the FAA Modernization and Reform Act provided an alternate route companies could take to gain the agency's permission to fly. Section 333 represented a "categorical exemption" to federal aviation regulations, Becklund said. "You've got to wonder: what good are the test sites?" Once the FAA made known it was entertaining Section 333 exemptions, the number of applications quickly ballooned to one hundred and beyond. "It was very disconcerting for us because the larger companies, a number of them dropped off our plates and showed up in a 333 (application)," Mooney recalled. "Essentially companies have been going to lawyers in New York and in DC, especially, and they're paying to have the lawyer write up an exemption. The lawyer goes out and hires aeronautical and aviation people to write the exemption, and we were told that we were not allowed to do that initially. Customers took off. People we were talking to said 'OK, fine,' and they left." Eventually, the FAA's UAS Integration Office "conceded that it makes sense to let us be the consultants to help" and allowed the test sites to process Section 333 applications.

Soon after she arrived at Virginia Tech, Mooney suggested the test site directors hold biweekly teleconferences and the others agreed. They resolved to present a united front in their dealings with the FAA. "I had conversations with the other executive directors and told them about what we used to do with UNITE as companies to push the initiative," she said. "We all decided there was value in working the issues that we have under our contract with the FAA as one, so that we would be able to come with one voice."

Textron Systems started testing the catapult-launched Aerosonde Mk 4.7 for commercial applications in late 2014 near Blackstone, Virginia, through a certificate of authorization the FAA granted to the Mid-Atlantic Aviation Partnership. The aircraft, which was equipped with a transponder to report its position and weighed seventy-five pounds, flew in airspace managed by Allen Perkinson Airport. Textron acquired the Aerosonde through its acquisition of AAI Corporation in 2007. *Courtesy of Textron Systems*

The progress of the test sites, or the lack thereof, was a topic lawmakers raised during a House Aviation Subcommittee hearing on December 4, 2014. Peggy Gilligan, the FAA's associate administrator for aviation safety, represented the agency on a panel of expert witnesses the subcommittee called to testify. "We're getting reports that the test centers are somewhat frustrated because there's not the interaction that they were expecting and we're not getting results," the chairman, New Jersey Republican Frank LoBiondo, said, addressing Gilligan. Rep. Dina Titus, a Nevada Democrat, had heard much the same in her state. "The enthusiasm is starting to wane because that test site is not producing like we thought it would," Titus said. "They just don't think it's getting off the ground, so to speak." Gilligan answered the FAA had started participating in teleconferences with the test site directors and was considering other practices that could help them, such as granting "broad COA authorization" of more than one aircraft at a time. "I think we have begun to alleviate some of those early concerns," she said. "I do think the test sites got off to perhaps a slower start than we and they were anticipating, as they came to understand what it was they had undertaken."

Center of Excellence

Legislation that passed earlier that year—the Consolidated Appropriations Act of 2014—contained a provision that further involved the states in the research and development effort. Congress directed the FAA to establish a "center of excellence" for unmanned aircraft systems, a consortium that would marshal the resources of federal agencies, academic institutions, and industry to help solve the UAS integration challenge. Unlike the test site program the FMRA legislation launched, the center of excellence came with money attached: Congress appropriated $5 million for the effort over five years. The FAA would tap the funding to award research grants that consortium members would match.

Following a year-long "rigorous competition," the FAA announced on May 8, 2015, that a coalition of universities and companies led by Mississippi State University would be the agency's UAS center of excellence. MSU's Alliance for System Safety of UAS through Research Excellence (ASSURE)—the center operator—combined fifteen research universities and other associate members. ASSURE members were already involved with three FAA test sites, had seven airfields, and operated 340 unmanned aircraft—more than the US Air Force, the consortium declared. The center's work would be dispersed among its members, but would be concentrated at NASA's John C. Stennis Space Center, near the border with Louisiana, to take advantage of airspace over the Gulf of Mexico, in the Mississippi Delta to conduct "precision agriculture" research, and at MSU's Raspet Flight Research Lab in Starkville, Mississippi.

CHAPTER 10

THE INDUSTRY EMERGES

In December 2013, the CBS news program *60 Minutes* broadcast an interview with Amazon CEO Jeff Bezos, during which Bezos revealed the online retailer's ambition to fly small packages to customers' doorsteps using drones. In fact, Amazon was already engaged in research and development of its planned "Prime Air" delivery service, said Bezos. With correspondent Charlie Rose in tow, he led a tour of the company's Seattle headquarters that opened into a room in which several octocopters were arrayed on a table. In concert with the televised announcement, Amazon's website featured a video showing an employee placing a small box into a plastic container that snapped shut. The container moved by roller conveyer to a waiting octocopter with its rotors already spinning. A clamp mechanism on the aircraft's belly grabbed the container; the drone then lifted off through an open warehouse door, crossed a field, and deposited the container at the doorstep of a waiting customer.

"I know this looks like science fiction; it's not," Bezos told the visibly enthralled Rose. Prime Air octocopters flying autonomously to programmed GPS waypoints could deliver packages of up to five pounds within ten miles of a fulfillment center. "This is early. This is still years away," Bezos cautioned. "The hard part here is putting in all the redundancy, all the reliability, all the systems you need to say, look, this thing can't land on somebody's head while they're walking around their neighborhood. I don't want anybody to think this is just around the corner. This is years of additional work from this point." Pressed by Rose, he surmised that 2015 would be the earliest possible date Amazon could introduce drone delivery service, given the FAA's mandate from Congress to open up the nation's airspace to drones. "My guess is that's probably a little optimistic," he added. "But could it be four, five years? I think so. It will work and it will happen and it's going to be a lot of fun."[1]

Though the industry was quite mature by then—AUVSI's annual conference that year in Washington, DC, drew 593 exhibitors, 8,000 registered attendees, and a demonstration by Code Pink—the Bezos announcement on *60 Minutes* was an epiphany of sorts for the American public. To use an overused word from the time, it went "viral" in other media accounts. The announcement "has created a media storm, and seems to have taken most people by surprise," Mary Louise "Missy" Cummings, director of the Humans and Automation Laboratory at the Massachusetts Institute of Technology, opined in *The Boston Globe*. "Many thought this was simply just a trick to boost

(Amazon's) sales for Cyber Monday, and is not a near-term reality. They could not be more wrong…Unless we address the regulatory aspects soon, we will watch the imminent commercial drone revolution happen overseas. Congress needs to hold the FAA's feet to the fire before this technology takes flight and leaves the U.S. commercial market behind."[2] Of note, the German company Microdrones flew its md4-1000 "Paketcopter" across the Rhine River to deliver a package of medicine to DHL's Bonn headquarters that same month. Deutsche Post DHL was also exploring drones.[3]

Not everyone was impressed by Amazon's prime-time revelation. In a column for the online magazine *Slate*, Konstantin Kakaes derided the *60 Minutes* broadcast as an "infomercial" and questioned Bezos's contention that delivery by drone was on the horizon. "Many media outlets have credulously repeated this claim, just like they did with the beer-delivering drone and the taco-delivering drone," Kakaes wrote. "However, the technical, regulatory and logistical challenges of autonomous flight in crowded American urban airspace are far more profound than Bezos allowed on TV…If thousands of drones are to fly around delivering packages across cities, they must become orders of magnitude more reliable than they are."[4] Meanwhile, the FAA had tempered its rolling estimate of the coming commercial market for small drones. The agency's annual aerospace forecast in both 2011 and 2012 projected a fleet of 10,000 such aircraft within five years, subject to its own ability to roll out regulations. Its 2013 forecast scaled back the number to 7,500.[5]

After Amazon, more announcements followed from other Internet-age companies, not about miniature helicopters flying nap-of-the-earth in the neighborhood, but about high-flying, solar-powered planes that could serve as communications nodes in the stratosphere. In March 2014, social media site Facebook acquired Ascenta-Aerospace of Yeovil, UK, whose principals had worked on projects including the Breitling Orbiter balloon and early versions of the Qinetiq Zephyr, a solar-powered UAV that in 2010

Social media website Facebook revealed a prototype of its solar-powered Aquila unmanned aircraft in a video on July 30, 2015. The Aquila, which had a 137-foot wingspan, was assembled in Yeovil, England, following Facebook's March 2014 acquisition of Ascenta-Aerospace. *Courtesy Facebook*

set an endurance record for unmanned aircraft by staying aloft for 336 hours and 22 minutes. "We're looking at a new type of plane architecture that flies at roughly 20,000 meters," Facebook engineering director Yael Maguire explained in a YouTube video. "That's a point where the winds are the lowest, it's above commercial airlines, it's even above the weather and can actually stay in the air for months at a time. These planes are solar-powered and they sit there, and they just circle around, and they have the ability to broadcast Internet down, but significantly closer than a satellite does." Facebook's ambition was to close gaps in global Internet coverage by "free space optics," using lasers to distribute high-capacity data streams "through the air and between platforms."[6]

Less than a month after the Facebook announcement, Internet search engine and advertising company Google confirmed it had purchased Titan Aerospace of Moriarty, New Mexico, developer of the Solara "atmospheric satellite" and earlier rumored to be an acquisition target of Facebook. Google also shared Amazon's ambition to deliver small packages by drone. In the summer of 2014, the company revealed its secretive Google X research laboratory in Mountain View, California, had started exploring a delivery service more than two years earlier, an effort it called Project Wing, and it had already staged a series of flight tests at a farm in Australia, deploying a prototype flying wing to transport items including chocolate bars and cattle vaccine. To lead the effort, Google hired Dave Vos, a South African and MIT-educated technology entrepreneur who had earlier co-founded a company (Athena Technologies) that developed miniature flight control systems for unmanned aircraft. Rockwell Collins, a major avionics manufacturer, acquired Athena Technologies in 2008.

In July 2014, Amazon was among the first companies to apply for an exemption under Section 333 of the FAA Modernization and Reform Act to fly drones. Its plan was to conduct outdoor research and development of the Prime Air experimental fleet over land it owned near Seattle. The Internet retailer was not among the first companies awarded exemptions in September that year, all of them aerial video companies associated with Hollywood filmmakers. Nor was it included in the second round of exemptions the FAA awarded in December. Among those companies were Trimble Navigation of Sunnyvale, California, which planned to operate its UX5 flying wing for aerial surveying purposes, and an Oregon company that planned to use the Aeryon SkyRanger quadcopter for flare stack inspections of Shell Oil platforms in the Gulf of Mexico.[7] Instead, the FAA requested that Amazon provide additional information to support its application.

In a supplemental letter addressed to Jim Williams of the FAA's UAS Integration Office in December, Amazon vice president for global policy Paul Misener vented frustration over the delayed exemption and suggested Amazon might take its drone ambitions offshore. The company was already eyeing India as a Prime Air launch country, according to media reports.[8] "Given the very conservative nature of Amazon's request, especially in contrast to the nature of other petitions for commercial small UAS operations that already have been granted, as well as the way that thousands of hobbyists are permitted to operate small UAS today, we are very concerned that our needs for testing operations have not yet been accommodated," the Misener letter stated. "... Subject to demonstrating the appropriate safeguards that Congress outlined in Section 333, Amazon should be free to conduct research and development operations over our own private property in the United States without further delay. Without approval of our testing in the United States, we will be forced to continue expanding our Prime Air R&D footprint abroad."[9]

Testifying before the House Aviation Subcommittee, the FAA's associate administrator for aviation safety, Peggy Gilligan, said the agency was trying to steer Amazon toward obtaining a special airworthiness certificate for Prime Air's multi-rotor helicopter, in addition to the Section 333 exemption. "We've been working with that applicant," she said, without naming Amazon. "What we are recommending is that they seek certification for their vehicle under our special certification rules for the purposes of research, and we think we can actually enable them to accomplish what they need to accomplish here in the US through the test sites and through their own certification." She added: "We know that they are not satisfied that they have to go that path."

Google and Amazon may have been darlings of California's high-tech Silicon Valley and the Pacific Northwest, but when it came to drones and influencing FAA policy, they evidently felt the need for a greater presence in Washington, DC. In October 2014, the cagey Internet giants were named among founding members of the new Small UAV Coalition with 3D Robotics, Airware, and camera manufacturer GoPro—all California based—along with foreign firms DJI Innovations, Parrot, and Aerialtronics. Michael Drobac, a former legislative counsel and Expedia and Netflix executive who was then serving as a senior policy advisor with the law firm Akin Gump Strauss Hauer & Feld, was the coalition's founding executive director.

Ironically, an Australian startup company named Flirtey upstaged the Internet giants in what was hailed as the first FAA-approved drone delivery operation on July 17, 2015. The Virginia Tech-led Mid-Atlantic Aviation Partnership managed the demonstration, which took place in the coal-mining region of southwest Virginia. NASA Langley Research Center contributed by flying its optionally piloted Cirrus SR22 single-engine airplane by remote ground station from Tazewell County Airport to Lonesome Pine Airport, some thirty-five nautical miles to the west. The SR22, with a safety pilot aboard, came bearing medical supplies. From Lonesome Pine Airport, a Flirtey pilot flew the pharmaceuticals in batches to a medical clinic at nearby Wise County Fairgrounds using a hexacopter. The six-rotor drone lowered the medicine to the ground by tether and returned to base to resupply. Virginia Governor Terry McAuliffe was among spectators at the event, dubbed a history making "Kitty Hawk moment" by its organizers. A veteran of similar demonstrations in Australia and New Zealand, Flirtey had earlier allied with the University of Nevada-Reno to test its drones at the university's indoor flight facilities.

The media buzz the Internet-age companies generated with their drone ambitions overshadowed an announcement that Lockheed Martin and AeroVironment made in early 2014 at the relatively obscure venue of the Defense Expo conference in New Delhi, India. The aerospace companies had joined forces to advance AeroVironment's liquid hydrogen-fueled Global Observer UAS as an atmospheric satellite for communications relay, storm tracking, and surveillance.[10] The Global Observer started life under a joint capability technology demonstration the Departments of Defense and Homeland Security awarded to AeroVironment in 2007. In April 2011, one of the two planes under development crashed on its ninth test flight. AeroVironment acquired the second plane from the program and continued its development. "We believe Global Observer can become a major growth opportunity for AV in three to five years, and we continue to work with multiple customers who are interested in this unique capability," AeroVironment chairman and CEO Tim Conver related during an earnings conference call in November 2013.

Missy Cummings had escorted *60 Minutes* correspondent Morley Safer on a walk-through of AUVSI's 2013 exhibition in Washington, DC, for a segment that aired seven months later in March 2014. "It may surprise you to learn that drones are flying across America," Safer intoned at the start of the episode. "Not the impersonal killing machines that patrol the badlands overseas, but drones nonetheless. Used by the FBI, by university researchers, by amateur photographers, even by your nosy neighbors. Domestic drones are poised to become a multibillion-dollar industry, revolutionizing everything from crop management down on the farm to, possibly, package delivery to your doorstep."[11]

This was not news to the industry, which had been making that case for years already. But Amazon delivering packages to one's doorstep was something the public could understand. "Whether Bezos got it right or not about the ability of Amazon to deliver in five years is not quite the point," AIA president and CEO Marion Blakey told journalists and company flacks gathered for AIA's 2013 year-end luncheon in Washington, DC. "It did give everyone a sense of the kind of beneficial applications that UAS will bring. We expect therefore that the FAA will take the hard but necessary steps to make that a reality."

Growth Industry

UAVs shaped as airplanes or helicopters could do many things. Package delivery and atmospheric satellites were among the edgier concepts. Real-life applications were within reach or already progressing for agricultural spraying and aerial remote sensing, law enforcement, search and rescue, fire fighting, wildlife monitoring, weather forecasting, newsgathering, power and pipeline patrol, industrial flare stack inspection, mining, surveying—the list went on. Advocates liked to talk about the four "Ds" when describing the utility of UAVs—the dull, dangerous, difficult, or dirty jobs that were either too risky, too expensive, or otherwise impractical using manned aircraft. "This industry very much reminds me of the energy that was around the Internet in the very early days," mused entrepreneur Brad Hayden, president and CEO of a business called Robotic Skies. "I think that we're going to see an explosion in the utility of these craft." Hayden was walking the show floor at AUVSI's 2014 conference in Orlando to advance his vision of a nationwide network of FAA certified aviation repair stations specializing in drones. "We're not talking the hobby stuff," he clarified. "Once you start getting into the tens of thousands of dollars on an airframe, and if you're using it commercially, where are you going to get that fixed? The hobby store?"

In the 2014 release of its "World Unmanned Aerial Vehicle Systems" market study, the respected consulting firm Teal Group said the market for UAVs outside of military uses promised to be one of aerospace's fastest growing segments over the next decade, but its growth, while "explosive," would spring from a low base. Teal estimated the civil/commercial market would amount to some $1 billion annually in 2023, less than 10% of the overall UAV market in value. Non-military government use of unmanned aircraft by agencies such as the US Coast Guard and Customs and Border Protection would be the largest single portion of the market due to the government's easier access to the airspace. The commercial side would be founded on mini and small UAVs—more substantial in numbers but lesser in value than larger, costlier machines. "A likely scenario" was that government agencies would drive growth by acquiring quantities of UAVs, as would companies that leased them to government agencies and other users who needed them for farming and other purposes.[12]

AUVSI released a broader, economic impact study in March 2013, that concluded the most promising commercial and civil markets for UAVs, respectively, would be "precision agriculture" and public safety, with the latter market encompassing law enforcement, fire fighting, and emergency response. The industry study, authored by airline analyst, aviation consultant, and former George Washington University professor Darryl Jenkins and Embry-Riddle Aeronautical University professor and aviation consultant Bijan Vasigh, predicted introducing unmanned aircraft into the national airspace system by 2015 as Congress directed would generate $13.6 billion in economic value in the first three years. Considering the number of sales, the demand for supplies, the number of jobs created, and tax revenue to the states, the industry would generate $82.1 billion between 2015 and 2025. The states gaining the most in terms of job creation and tax revenue would be, in descending order: California, Washington, Texas, Florida, Arizona, Connecticut, Kansas, Virginia, New York, and Pennsylvania. By 2025, the industry would create 103,776 total jobs. Every year that integration was delayed, the study found, the country would be denied $10 billion in unrequited economic impact. "Most of the barriers of potential usage are governmental and regulatory," the authors wrote. In the absence of regulatory guidelines from the FAA allowing UAVs wider access to the airspace, they added, "this report is simply the opportunity cost to the economy of a good idea that was hindered due to government interference or inaction."[13]

Commercial agriculture represented $75.6 billion of the projected $82.1 billion in total economic impact over the next decade, dwarfing other markets. The authors assumed the adoption rate of unmanned aircraft technology by the farming industry in the United States would mimic that of Japan, where the market experienced early growth rates in excess of 20% annually from a zero base in 1990 before growth tapered off.

Precision Farming

The anticipated market for unmanned aerial farming was bifurcated into two segments: remote sensing using cameras to monitor crops for outbreaks of disease, hydration, or growth rates; and precision application of herbicides, pesticides, fertilizers, water, and seeds. The former favored smaller, lighter aircraft that could perform predefined sweeps over fields or hover above crops, ranging from models such as the three-pound PrecisionHawk Lancaster Mark III, which

Baptiste Tripard, business development manager with Swiss company senseFly, shows off the eBee flying wing at the Unmanned Systems 2014 conference in Orlando. Made of foam and carbon fiber, the 1.5-pound eBee carries a near-infrared camera or other sensors for crop monitoring. *Author photo.*

The PrecisionHawk Lancaster Hawkeye Mark III is shown against a backdrop of red cabbage and beets. In addition to the unmanned airplane, the company, founded in Toronto by veterans of the Canadian Space Agency, developed a Low-Altitude Tracking and Avoidance System for transponder-equipped aircraft. *Courtesy PrecisionHawk Inc*

resembled a model airplane that could land on a lake and take water samples when fitted with floats; to flying wings such as the AgEagle, senseFly "eBee," or Trimble UX5; to multirotor helicopters such as the Lockheed Martin Indago, Falcon Hover, or 3D Robotics X8-M. There were many more different aircraft types. A Section 333 exemption the FAA granted to unmanned aircraft services company Measure in the summer of 2015 listed 324 different small UAS it could operate for "aerial data acquistion" jobs under certain conditions.

In July 2015, the American Farm Bureau Federation unveiled an "ROI Calculator" it had developed with Measure and consulting firm Informa Economics. Described as the first such web-based tool available to farmers, the calculator estimated the return on investment from using small drones for crop scouting, 3-D terrain mapping, and crop damage assessment by insurers. It initially covered corn, wheat, and soybeans, three of the largest production crops. In the case of crop scouting—using a drone service to collect visual and near infrared imagery—the calculator found the average US farmer would realize a return on investment of $12 per acre for corn, $2.60 for soybeans, and $2.30 for wheat. Principal sponsors of the calculator effort included Lockheed Martin and food and beverage giant PepsiCo.[14]

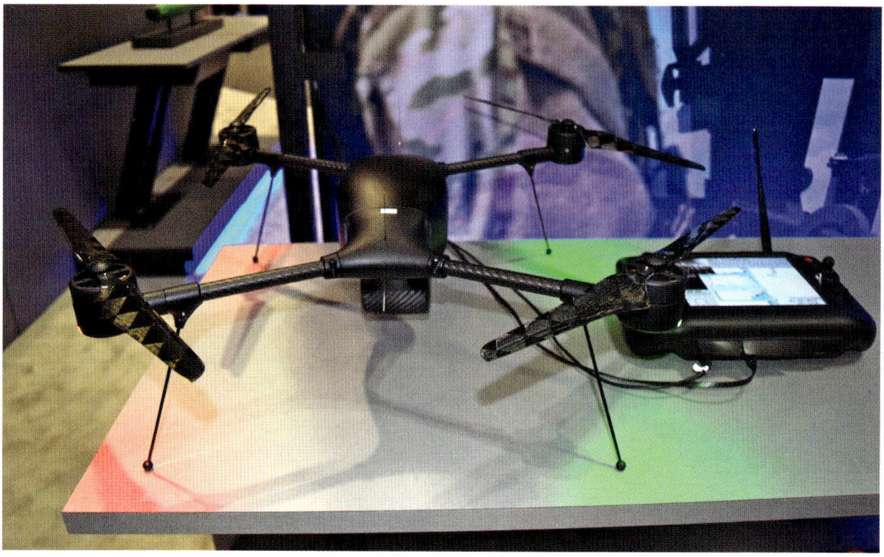

Among companies vying for the emerging market for unmanned aerial farming was Lockheed Martin, with its five-pound Indago quadcopter. The Indago resulted from the defense contractor's acquisition of miniature autopilot developer Procerus Technologies in January 2012. *Author photo.*

Drone system provider Elevated Horizons said its Agri6 hexacopter, fitted with a thermal imaging camera, helped detect laurel wilt disease threatening avocado crops in Florida. Produce giant Dole used the Agri6 in Costa Rica. *Author photo.*

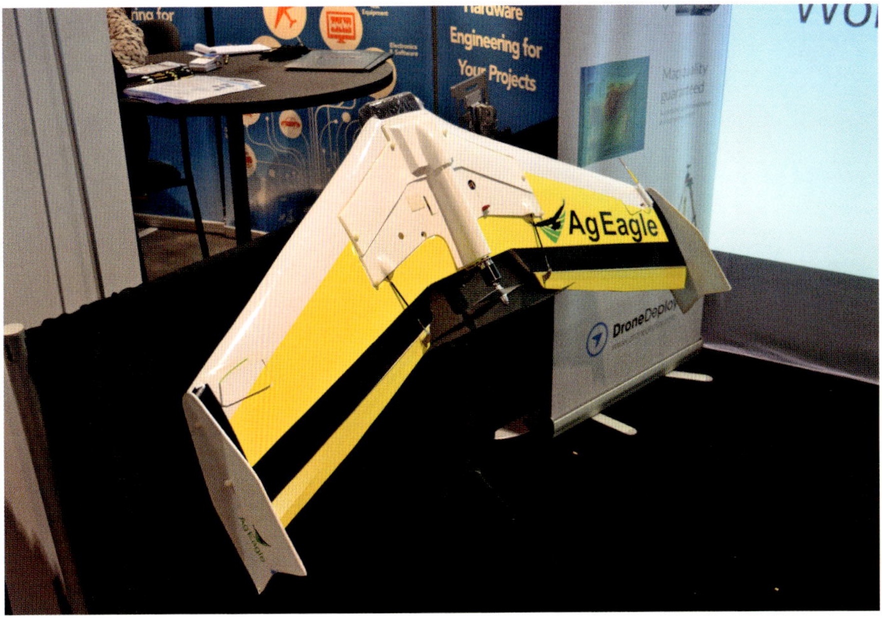

AgEagle LLC of Neodesha, Kansas, produced a flying wing by that name for precision agriculture. The AgEagle, displayed at the Unmanned Systems 2015 conference, was based on technology developed by the Kansas State University agronomy department. *Author photo.*

Contending for the application segment was Yamaha Motor Corporation USA with the remotely piloted RMAX helicopter, capable of laying down 4.25 gallons of liquid or seven gallons of granular spray from tanks or hoppers mounted on either side of its fuselage. Based in Cypress, California, the company was a wholly owned subsidiary of Japanese motorsports manufacturer Yamaha Motor Co. In 1983, Japan's Agricultural Aviation Association commissioned Yamaha to develop a remote-controlled aircraft capable of spraying agrichemicals on rice paddies and other crops. After enlisting the cooperation of Japan's top maker of model helicopters, Yamaha started producing the R-50 unmanned helicopter in 1987. This was succeeded by the improved RMAX series in 1997. By 2013, there were 2,458 registered Yamaha helicopters operating for agricultural purposes in Japan. In terms of rice alone, they treated more than one million hectares, or some 35% of the country's rice paddy fields. "By simple calculation, this means that one in every three bowls of rice served in Japanese homes has been grown with the agrichemical pest control spread by a Yamaha unmanned helicopter," the company declared.[15] Yamaha also supplied some 150 RMAX helicopters over a decade to South Korea, and in 2011 introduced the RMAX in Australia, where it was used for weed control on ranches, mining sites, dam walls, and powerline corridors.

The FAA granted Yamaha's US subsidiary and the University of California-Davis a COA in November 2012 to test the RMAX at the university's forty-acre Oakville Experimental Vinyard set in the Napa Valley wine growing region. The lead researcher on the project, agricultural engineering professor Ken Giles, and development engineer

Steve Markofski, an executive with Yamaha Motor Corporation USA, answered questions about the RMAX unmanned agricultural helicopter the company displayed at the 2014 AUVSI conference in Orlando. The Japanese company's US subsidiary hoped to supply it to farmers for aerial spraying of crops. *Author photo*.

Ryan Billing set out to build "a valuable database that will document how the RMAX might perform in agricultural operations in California and elsewhere in the United States," reported *Vine Lines*, a UC Cooperative Extension newsletter. "So far, the data indicate that the helicopter is providing thorough coverage across the vineyard and that the air currents stirred up by the helicopter rotors are causing the spray to reach even the undersides of the grapevine leaf canopy. Furthermore, the researchers have been impressed by the stability of the helicopter, even in gusty winds."[16] The UC Davis project built confidence in the helicopter, said Steve Markofski, Yamaha USA's executive for new business planning. "That COA has evolved from only allowing us to spray water and doing simple deposition tests to actually treating a field with chemicals," he reported. "They treat the field with their ground rigs and we have a portion of the vineyard that is ours to treat that's roped off. We come back the next day and treat it with the RMAX."

The Napa Valley testing showed the RMAX could treat up to eleven acres of vineyard in the same time a tractor took to cover 1.5 acres, the manufacturer asserted in testimony to the Senate Transportation Committee. "The RMAX uses a fraction of the fuel, causes no soil compaction or crop damage, and provides more precise spray deposition, significantly reducing chemical drift and operator exposure to the chemicals," stated Henio Arcangeli, the US subsidiary's vice president of corporate planning.[17] Testers also deployed the 207-pound takeoff weight helicopter at Paramount Farms, outside Bakersfield, California, which described itself as the world's largest grower and processor of almonds and pistachios. The RMAX showed it would be "ideal" for treating crops against navel orangeworms that threatened the industry, Arcangeli told senators. Although the helicopter was manufactured in Japan, Yamaha was open to producing RMAX units in the United States, potentially creating hundreds of jobs, he testified.

Once the RMAX gained wider acceptance in the United States, the business model would likely approximate that in Australia, where Yamaha leased the helicopter to spray operators and retained ownership of the machines, Markofski said. "Yamaha would be responsible for maintenance of the aircraft and training the operator," he explained. "What it also affords us is a greater degree of control. We know who has the aircraft, where it's being operated, how it's being operated." The helicopter could also be "geographically locked" or geo-fenced, to only operate within a defined area.

Yamaha displayed the RMAX at AUVSI's 2014 Unmanned Systems conference in Orlando. Nearby for the first time in the United States was its successor—the Fazer—featuring 50% more payload capacity. And not far away, Northrop Grumman displayed the new "R-Bat," an autonomous version of the RMAX, painted in military gray, that the companies had just unveiled for reconnaissance, search and rescue, border security, and other missions. This was not the first time Yamaha displayed the RMAX at the annual industry gathering, but this was the helicopter's first showing in several years. The regulatory landscape in the United States had shifted in the interim, hastened by the FMRA legislation. Yamaha sought to introduce the RMAX along two tracks: the company had approached the FAA's Los Angeles certification office to explore the type certification process, and it also planned to seek an exemption to operate the RMAX under Section 333 of the legislation. "What I think it will do…is educate everyone on who you are and who you're not," Markofski said. "Obviously we want to begin commercial operations, but at the same time it's a great opportunity to prove ourselves."

The WATT-200 developed by Drone Aviation Holding Corp., of Jacksonville, Florida, was designed to provide secure and reliable aerial monitoring for extended periods of time while tethered to the ground with a high-strength, armored tether. The Mid-Atlantic Aviation Partnership announced in August 2015 it would collaborate with the company to advance the technology. *Courtesy Drone Aviation Corp.*

Aerovel, a company headed by Insitu co-founder Tad McGeer, developed the Flexrotor, seen here at the Unmanned Systems 2015 conference in Atlanta. Weighing forty-five pounds, the Flexrotor is a "tailsitter" aircraft that transitions from vertical thrust-borne to horizontal wing-borne flight. *Author photo*.

Markets other than agriculture held promise for unmanned aircraft. A 2013 paper the American Institute of Aeronautics and Astronautics sponsored analyzed the savings operators might realize by replacing manned aircraft with drones for three specific missions: pipeline patrol, law enforcement, and mid-sized cargo delivery. "Throughout the analysis, we differentiated the missions and estimates of market size for the various UAS applications, ultimately determining that UAS technology can provide significant cost savings when compared to the status quo," stated the authors: Benjamin Litvinas, an economist then with the consulting firm GRA; and Nathan Dickerson, of Booz Allen Hamilton.

According to the AIAA paper, small piston airplanes and helicopters regularly patrolled some 400,000 miles of petroleum and natural gas pipeline in the United States as required by the Department of Transportation's Pipeline and Hazardous Materials Safety Administration. Interviews with subject matter experts revealed safety was a major consideration. Flying at about 500 feet above the ground to check for abnormalities, pilots risked colliding with guy wires, radio towers, and terrain along pipeline rights-of-way. The authors matched a hypothetical fleet of single-engine Cessna airplanes and Robinson R22 and R44 helicopters—ranging in cost from $1.50 to $5.50 per mile on average to operate—against an Insitu ScanEagle costing $1.67 per mile. Assuming the ScanEagle eventually penetrated 25% of the market, they concluded that unmanned aircraft would save $42.4 million over the period 2015 to 2017, when discounted for inflation. The costs to society of an estimated 0.87 non-fatal and 1.349 fatal accidents per year attributed to pipeline patrol would be eliminated.

For the law enforcement mission analysis, the authors replaced Cessna and Beechcraft airplanes with the hand-launched AeroVironment Raven and Bell and

CyPhyWorks built the Persistent Aerial Reconnaissance and Communication (PARC) tethered drone that made use of the company's patented microfilament tether technology to support long duration flights with secure streaming of high-speed data and HD video. Interested in the PARC system, Motorola Solutions made a strategic investment in CyPhyWorks in March 2015. *Author photo*.

Airbus helicopters with the Honeywell T-Hawk, a vertical take-off and landing machine propelled by a ducted fan. They estimated average hourly operating costs of $820 for an airplane, including the costs of fuel, pilot, and maintenance, versus $140 for the Raven; and $2,400 for a helicopter versus $140 for the T-Hawk. Again, assuming 25% market penetration by the drones, they concluded police would save $725 million from 2015 to 2027, which agencies would likely "recycle" into additional flight hours. For the cargo market analysis, they considered the savings that might result from replacing FedEx's single-engine Cessna 208 Caravan turboprops with an unmanned version of the plane, but they acknowledged that FedEx had no plans at the time to do that.[18]

A survey the National Press Photographers Association conducted measured the news media's interest in using small drones for newsgathering purposes. The NPPA was among the news media *amici* that aligned with Raphael Pirker in his legal battle against the FAA on First Amendment grounds. So too, for that matter, was *The Washington Post*, which Amazon's Jeff Bezos purchased in September 2013 for $250 million. Emailed to fifty news organizations and media associations between February and mid-March of 2014, the survey garnered 680 responses, mainly from individuals who identified themselves as journalists, but also from news managers, attorneys, academicians, and students. "Although there were a statistically valid number of responses those

CNN chief editorial counsel David Vigilante explains the cable network's interest in using drones for news gathering at the Unmanned Systems 2015 conference in Atlanta. Accompanying him on stage were (l-r) Colin Guinn, 3D Robotics chief revenue officer; CyPhy Works CEO Helen Greiner; and Dave Vos, leader of the Google X Project Wing effort. *Author photo*.

answers should not be inferred to the industry at large; but while respondents were self-selected their anecdotal answers are very instructive," NPPA general counsel Mickey Osterreicher advised in a paper analyzing the results.[19]

The survey asked whether journalists and other media professionals would be interested in using drones for newsgathering "if it were legally permissible and within a reasonable budget." Of the 559 respondents to that question, 62% said they would be most interested. Eighty-six percent of 570 respondents believed journalists had a First Amendment right to use drones. Asked how drones could be used for newsgathering, 302 people answered. "Many wrote (that) it was a better way to cover major news events such as accidents, fires, natural disasters, and sports. Some referenced access to difficult terrain combined with the ability to provide overall views and unique perspectives not normally seen. Others cited cost compared to a helicopter as well as the safety factor in using an unmanned aircraft."

Osterreicher noted that a majority of survey respondents stated they worked outside of the television industry, which traditionally was the only sector that could afford newsgathering helicopters. Interest in small drones by the "non-TV cohorts" suggested the competitive advantage of larger broadcasters could be challenged by smaller outlets deploying unmanned aircraft. The paper concluded with some recommendations, among them the need for "reasonable regulations" governing the use of drones for newsgathering. "Requiring proof of flight proficiency, a vehicle registration scheme and minimum insurance, similar to motor vehicle operation make sense," it stated. "A pilot certification course for sUAS that is far more costly than the devices themselves, along with burdensome reporting processes and onerous requirements for filing flight plans are counter-intuitive and will be counterproductive."

Small drones may have been the poor man's solution to electronic newsgathering, but large broadcasters were not disinterested in the technology. In June 2014, cable network CNN announced it had entered into a joint research initiative with Georgia Institute of Technology to study media use of drones, an agreement that was later expanded to include cooperative research with the FAA. In March 2015, the network featured aerial footage it captured with a drone of the Edmund Pettus Bridge and the cityscape of Selma, Alabama, for the 50th anniversary of the "Bloody Sunday" civil rights march. It conducted the shoot at dawn under an FAA certificate of authorization it obtained with Georgia Tech using a DJI Spreading Wings S1000 multicopter equipped with a Panasonic video camera. HeliVideo Productions, of Austin, Texas, operated the aircraft.[20] CNN later figured in the highest profile news from the Unmanned Systems 2015 conference. Appearing at the industry event in Atlanta, FAA Administrator Michael Huerta announced that CNN, BNSF Railway, and drone manufacturer PrecisionHawk would participate with the agency in a "pathfinder project" to explore using small fixed-wing and multi-rotor drones in various scenarios. BNSF partnered with Insitu to conduct the initiative's first beyond-line-of-sight flight on October 25, 2015—also the first such commercial operation in the lower forty-eight states. They launched Insitu's *ScanEagle* from a site near Vaughn, NM, to provide video surveillance along a sixty-four-mile length of track, about half of the section BNSF designated for a weeklong exercise. CNN's contribution was to research how drones might be deployed for visual line-of-sight operations in urban areas. Its initial plans involved two aircraft platforms: the AirRobot AR180 quadcopter and the Drone Aviation WATT-200 tethered drone.

ArrowData, a Las Vegas-based "aerial data services" company, laid claim to receiving the first FAA exemption to operate drones for newsgathering under Section 333 in April 2015. The company used a FreeFly CineStar X8 multi-copter equipped with a Sony video camera and TVU Networks mobile cellular video transmitter. Among early operations, it helped the ABC affiliate in San Francisco cover the demolition of historic Candlestick Park, becoming "DroneView7" for the mission. Former sportscaster Ron Futrell, ArrowData's director of marketing and media relations, coined the term "aerojournalism" to describe the emerging industry segment. "It's a little ahead of its time in the sense that newsrooms still have questions and are still wondering how best to do this," Futrell said during an interview at ArrowData's offices that summer. "Some of them are tentative, worried, concerned. They want to make sure to do it right, and that's a good thing. But at the same time, we've shown we can do it safely and responsibly and also enhance stories by using aerojournalism." He envisioned a time when broadcasters not only used small drones to cover current events, but even Predator-sized unmanned aircraft that could be moved to regional hotspots. "Absolutely I can see that," Futrell declared, although he conceded that such operations were years away at best. "Every TV station in the country, just as they have cameras, is going to have an aircraft to fly for news."

ArrowData director of operations Bill "Mario" Wilson adjusts a CineStar X8 multi-rotor helicopter at the company's Las Vegas headquarters. One of the aircraft's supporting legs was a temporary fix. In April 2015, ArrowData received the first FAA exemption to fly drones for electronic news gathering and coined the term "aerojournalism" to describe the new segment. *Author photo.*

Insurance Considerations

At least for the near term unmanned aircraft did not carry passengers, but their operators still faced exposure to third-party bodily injury and property damage, much like the operators of manned aircraft. "[T]he potential for a liability claim being made against a UAS operator in the event of an aircraft accident is real and should not be dismissed," advised Carson Lyons, vice president for aerospace with insurance firm AIG Property Casualty. Operators also needed to consider covering damage to the aircraft or to the equipment it carried, either of which could be expensive for a small business. Meanwhile, drone manufacturers faced potential liability for the sale and use of their products "if it were determined that something related to the design, manufacturing or maintenance of a particular aircraft or aircraft component was a factor in an individual or series of accidents or incidents," Lyons said.

In a written response to a set of questions from the author, Lyons said the insurance market for small UAVs had accelerated in the prior year from a few inquiries per month regarding the availability of coverage or requests for coverage to numerous such inquiries per day. Insurance carriers, including those involved in traditional aviation markets, standard commercial general liability carriers, and carriers of personal lines of insurance, were considering whether they wanted to participate in the drone market and, if so, how they should best underwrite individual operations, he said. "In general, the insurance industry has responded fairly quickly to what is truly a new type of exposure, however, as the numbers of small UAS increase, commercial operations grow and the inevitable losses occur, the breadth of market interest, coverage and underwriting standards will continue to evolve," Lyons wrote. "Today, many agents, brokers and underwriters see a good deal of potential in this industry which means good availability of coverage for small UAS physical damage and liability exposures." The cost of an insurance premium for using a drone depended on the individual operator, the type of operation, and his or her specific coverage needs. "[B]ut, depending on these numerous factors, pricing may be similar to what it costs to insure the family car, or run well into the tens of thousands of dollars."

Safety Record

Given the nascent state of the commercial industry, insurers lacked a safety record on which to assess small UAVs, and any comparison with military machines was not favorable. In June 2014, *The Washington Post* published the findings of a year-long investigation it conducted of worldwide crashes involving US military unmanned aircraft. The newspaper derived its information from more than 50,000 accident investigation reports and other records obtained through some two dozen FOIA filings with the four US military services. Taken together, the records revealed 418 major unmanned aircraft crashes between September 2001 and December 2013, due to mechanical malfunction, pilot error, lost C2 link, and other causes. There were forty-seven major crashes in the United States. The overall number was "almost equivalent" to the number of major crashes involving Air Force fighters and attack planes during the same period, although drones flew far fewer missions and hours. There were no drone-accident fatalities, but records showed that many potential catastrophes were narrowly averted. This was "a record of calamity that exposes the potential dangers of throwing open American skies to drone traffic," the *Post* stated in its lead sentence. The newspaper's investigation uncovered scores of previously unreported crashes, "challenging the federal government's assurances that drones will be able to fly safely over populated areas and in the same airspace as passenger planes."[21]

Specifically, the *Post* found the military had categorized 194 crashes as Class A accidents that either destroyed the aircraft or caused at least $2 million in damage. Just over half of those accidents took place in the war zones of Afghanistan and Iraq; nearly a quarter happened in the United States. The military categorized 224 crashes as Class B accidents that cost between $500,000 and $2 million. Among the most crash-prone models was the MQ-1 Predator. Almost half of the Air Force's 269 Predators were involved in a major accident, said the *Post*. However, the Predator's safety record had improved from 13.7 Class A accidents for every 100,000 flight hours during its first dozen years of existence to 4.79 since 2009. Its successor—the MQ-9 Reaper—in the previous five years had incurred 3.17 Class A accidents per 100,000 flight hours, but of late was approaching the crash rate of F-16s (1.96) and F-15s (1.47). Some 55% of the Army's MQ-5 Hunters and 38% of its RQ-7 Shadows had been involved in major accidents. Unavailable was information on the fleet of about thirty Predators and Reapers the CIA flew using Air Force pilots, or on the secretive, stealthy Lockheed Martin RQ-170 Sentinel, one of which Iran captured in December 2011.

Alarming in tone, the three-part investigative series by a newspaper most of Congress read prompted an immediate response from the Aerospace Industries Association, which leaped to defend its member manufacturers. "Today's *Washington Post* story, 'When

Drones Fall from the Sky,' ignores critical factors regarding safety of unmanned aircraft systems in a misguided article that could frighten readers about the impending integration of UAS into the national airspace," the AIA declared. "[I]t's wrong to conflate the performance of far heavier systems, largely operated in combat zones and in more extreme weather and terrain conditions, with the kinds of limited operations FAA currently is contemplating in the United States… Moreover, the loss rates that major military UAS systems are experiencing compare very favorably to the loss rates of manned systems at the same point in their development cycle. Nothing leads us to believe unmanned aircraft will not be as safe or safer than manned aircraft over the long term."[22] AUVSI's Michael Toscano sent a letter to the *Post*, saying that the series "influences the data to draw incorrect conclusions about the domestic use of UAS. Many unmanned aircraft used by the military operate under different operational environments and in different situations than will apply domestically." The *Post* did not print Toscano's letter, AUVSI said.[23]

Missy Cummings, then representing the Duke University Humans and Autonomy Laboratory that had relocated from MIT, made a similar case to the Senate Transportation Committee in January 2014. A former Navy F/A-18 pilot who wrote of her experiences as a female aviator in the 2000 book *Hornet's Nest*, Cummings had come to the fore as an advocate of unmanned aircraft. When comparing accident rates per 100,000 flight hours—the industry standard—unmanned aircraft had a higher rate than other aircraft over the last twenty years, she told the senators. But that was an apples-to-oranges comparison, since manned aviation had more than 100 years to improve its safety record and had in place formal certification and inspection programs. For the first five years they operated, drones were more than twice as likely to have an accident compared to a manned aircraft; after fifteen years, they were only 25% more likely to crash. "If this dramatic improvement in safety continues, theoretically drone safety could be on par with that of commercial aviation in just 10 years," Cummings testified.[24]

Close Encounters

Malfunctions and mistakes subjected large unmanned aircraft to potential damage or destruction, but the emerging threat of midair collisions from a burgeoning universe of small, highly capable but affordable drones that enthusiasts purchased on-line or through consumer electronics stores increasingly demanded attention. An unannounced rotorcraft or some other small drone flying in the vicinity of an airport was at best an unwelcome distraction for pilots during the critical takeoff or landing phases; smashed against the windscreen, clipped by a propeller, or ingested by a turbine engine, it could be potentially deadly. The Airbus A320 that ditched in the Hudson River in the famed "Miracle on the Hudson" incident in January 2009 was brought down by a flock of Canadian geese while departing New York's LaGuardia Airport, vividly demonstrating what birds—a long-standing problem for aviation—could do to an airliner. Could a three-pound quadcopter made of plastic and composites, with an encased lithium polymer battery the size of one's hand, do something similar? "No accident is the result of a single event," advised Tim Canoll, a Delta Air Lines captain and former Navy pilot who took over as president of the Air Line Pilots Association in 2015. "We think of it as a chain of events—each link of the chain being an element that if removed would have prevented the accident. It's very hard to point to any single element that could cause an aircraft to come down. That being said, we always look at each element—whether it be weather, air traffic control, fatigue, aircraft design, unmanned aerial systems—as potentially adding to an accident. We're constantly looking for ways to mitigate the threat posed by that link in the chain, and that's the way we're looking at unmanned aerial systems. They are a link in a chain that could lead to an accident." Beyond airports, drones were also a potential hazard for low-flying police and emergency medical helicopters, as well as crop-skimming agricultural aircraft.

Guidelines for the recreational use of "model aircraft" expressed by Congress in the FAA Modernization and Reform Act of 2012 and advanced by the FAA, the Academy of Model Aeronautics, and industry associations through the "Know Before You Fly" campaign launched in late 2014, recommended that hobbyists fly their machines no higher than 400 feet above the ground and, absent prior notification, no closer than five miles from an airport. By 2014, what started as a drip of rogue drone incidents that made national news—for instance, the Phantom that flew over the East Harlem gas explosion and the Phantom that circled Seattle's Space Needle—grew into a torrent. Although it would not come to light until several weeks later, on March 22, the pilot of a US Airways CRJ200 regional airliner reported nearly colliding with what appeared to be a radio-controlled replica of an F-4 Phantom fighter flying at 2,300 feet near Tallahassee Regional Airport. In its three-part series on drone accidents that June, *The Washington Post* called out two sightings near major airports, both on May 29. One involved a pilot who reported seeing a "gull wing" type aircraft at 5,500 feet while approaching LaGuardia Airport. The other was reported by the pilots of two aircraft approaching Los Angeles International Airport. They described seeing a "trash can sized" drone or model aircraft at 6,500 feet. In fact, drones had flown dangerously close to airports or passenger aircraft fifteen times over the previous two years the newspaper said, citing FAA records.

That November, the agency released further records in response to media requests that revealed a surge in reported drone sightings. *The Post* raised the alarm in a front-page article. "The newly released incident reports…reveal that the FAA has been receiving a steady stream of near-collision alerts from airliners and private pilots," reporter Craig Whitlock wrote. The agency was "struggling to keep up with an influx of cheap drones that have flooded the market."[25]

The FAA provided the author with a spreadsheet that listed 194 "Non-COA UAS" events (incident reports) involving other than authorized drones from February through November 2014. They were reported in thirty-seven states and the District of Columbia; one took place over Canada. There were twenty-five events in California, nineteen in New York, and seventeen in Florida. Washington, DC, was the location of eight events. According to one narrative, police "reported arrest of individual stuck in tree at Freedom Plaza, intersection of 14th St., and Pennsylvania Ave., NW, after he climbed up the tree to retrieve his UAS that had crashed into the same tree." This happened a couple blocks from the White House; other events were reported at the Lincoln Memorial and on the grounds of the US Capitol. In nearby Virginia, a Porter Airlines Q400 twin turboprop flying at 2,800 feet on approach to Washington Dulles International Airport "came within 50 feet of what was reported as a black and silver drone with three circular propellers."

Not covered by the list was the infamous drone event of January 26, 2015. In the early morning hours, a reportedly drunk off-duty employee of the National Geospatial Intelligence Agency launched a DJI Phantom from the window of a downtown

Washington apartment. He thereafter lost control of the quadcopter, which crash-landed on the south lawn of the White House, triggering a lockdown of the president's residence and national introspection about its security. Little more than a month later, DJI released an update to the flight controller firmware of its Phantom 2 series machines. In line with Washington's status as a special flight rules area, the update "geo-fenced" the quadcopter based on GPS coordinates, preventing it from operating within a fifteen-mile radius of the White House.

In May 2015, the FAA launched a "No Drone Zone" campaign to inform the public that unauthorized drone flights were forbidden in the city of Washington and communities within a fifteen mile radius of Ronald Reagan Washington National Airport—a restriction on aircraft operations stemming from the September 11 terrorist attacks fourteen years earlier. The day after the FAA's announcement, the Secret Service detained a man who officers observed flying a Parrot Bebop over Lafayette Park, located across the street from the north face of the White House. That fall, the FAA doubled the No Drone Zone, saying it applied to a thirty-mile radius of Reagan Airport.

Senators Dianne Feinstein of California and Charles Schumer of New York introduced the Consumer Drone Safety Act in mid-June 2015. As proposed, the legislation would give the FAA more power to regulate hobby drones. In a nod to the model aircraft community, the regulation would apply only to drones that were operated outside of the guidelines and programming of a "nationwide community-based organization," such as the Academy of Model Aeronautics. When final, it would set a maximum altitude

CyPhyWorks, headed by iRobot co-founder Helen Greiner, launched the development of its LVL 1 consumer drone through a Kickstarter funding campaign. The company promoted the smartphone-controlled LVL 1 as "the first drone for absolutely everyone" because it took high-quality video with its forward-facing camera and was easy to use. *Author photo.*

3D Robotics displayed its first consumer offering—the Solo quadcopter—at the Unmanned Systems 2015 conference. Positioned to compete against DJI's Phantom 3, the Solo weighed less than five pounds, ran on open-source software, and retailed for $995 for the drone itself—more for a gimbal and GoPro camera. *Courtesy 3D Robotics*

that consumer drones could fly to and establish areas where flights were restricted because of the possibility of "unsafe interactions with manned aircraft," or the risk of damage to persons or property on the ground. The regulation would require manufacturers of consumer drones "whether through software or other technological means" to cap their attainable altitude and prevent them from flying near airports or within protected airspace. Through "sensors and software or other similar means," drones were to come equipped with a collision avoidance capability and the means to safely and autonomously land in the event of a lost communications link with the operator. They were to have registration numbers that pilots and air traffic controllers could identify and transponders to signal their position in space. "If we don't act now, it's only a matter of time before we have a tragedy on our hands," Feinstein stated in a press release announcing the bill. "Consumer drones are a new technology. They can fly thousands of feet in the air and jeopardize air travel, but the FAA can only regulate them if they are used for commercial purposes. That loophole must be closed."[26]

There were more troubles with drones. In July 2015, a Connecticut teenager posted a YouTube video showing a quadcopter with a semiautomatic pistol clamped to its nose firing rounds in a wooded area, though apparently no laws were violated. That summer, drones interfered with fire fighting aircraft during several wildfires in southern California, including one that burned cars along Interstate 15 east of Los Angeles, forcing incident commanders to temporarily suspend operations.[27] State lawmakers reacted by advancing a pair of bills: one would indemnify emergency workers who damaged a drone in the course of fire fighting operations; the other held that "knowingly, intentionally or recklessly" interfering with such operations using a drone was punishable by a $5,000 fine, six months in jail, or both. "California is a scorched, drought-dried tinderbox that is facing unprecedented fire risk right now," said state senator Ted Gaines, one of the sponsors. "We need every firefighter, every shovel, every bulldozer and air tanker working together to battle wildfires. If drones are getting in the way of that effort they need to be grounded immediately."[28]

But the phenomenon of nuisance drones would not be easily suppressed. Apparently prompted by new revelations brought to light by *The Washington Post*, over a ten-day period in August 2015, the FAA released new information that suggested the problem was enormous. In a front page article on August 11, Whitlock wrote that rogue drone operators were becoming "a national nuisance, invading sensitive airspace and private property—with the regulators of the nation's skies largely powerless to stop them." The next day, the FAA announced pilot reports of errant drones had nearly tripled over the previous year, from 238 sightings in all of 2014 to more than 650 as of August 9. That month, multiple airline pilots spotted drones at John F. Kennedy and Newark Liberty International airports, according to media accounts. "The FAA wants to send out a clear message that operating drones around airplanes and helicopters is dangerous and illegal," the agency sternly advised. "Unauthorized operators may be subject to stiff fines and criminal charges, including possible jail time."[29] On August 21, *The Post* published a front page article with the banner headline: "Rogue drones clogging U.S. airspace." In that article, which appeared on-line the night before, Whitlock wrote that reports of close calls between drones and pilots were already closer to 700 in number. The newspaper had obtained several hundred reports "from a government official who objected to the FAA's secrecy." Later that day, the FAA released a list of 765 reported drone sightings by pilots from November 13, 2014, through August 20.

In the midst of the outbreak, there were countervailing opinions on the threat drones posed to commercial aviation—one notably came from Jim Williams, by then retired from the FAA and working for a major law firm. Quoted in *The Wall Street Journal*, Williams explained that aircraft were designed to absorb impacts from similarly sized birds and keep flying. "People tend to forget how big the sky is and how small these little quadcopters are," Williams said, according to the newspaper. "[T]he likelihood of one of these little guys taking down an aircraft is very remote."[30] Airline pilot Chris Manno, in his *JetHead* blog, said the media tempest over drones was much ado about nothing. "The hazard presented by unwanted objects in an aircraft's flight path is nothing new," Manno wrote. "In fact, each year hundreds of bird strikes are dutifully and without fanfare reported by airline pilots as required by law. What's new is the opportunity for media and aviation 'pundits' to claim more screaming headlines by overstating the drone hazard."[31]

Nevertheless, with the FAA still developing its small UAS regulation for commercial operations, there was mounting pressure on the agency from the air transportation industry to better control recreational drones. "Sooner or later—and I personally believe it will be sooner—an airliner full of passengers and crew is going to be brought down after colliding with a drone. There, I've said it, though most in the industry won't," wrote Karen Walker, editor of the airline industry trade publication *Air Transport World* in a blog post. "We must have an urgent, honest discussion about what is happening in the skies today. Even more urgent, we must do something about the rapidly escalating danger that drones pose to commercial air transport."[32] The headline of another blog post by Edward Wytkind, president of the Transportation Trades Department, AFL-CIO, announced: "We said go slow on drones. We weren't kidding." The TTD was a labor coalition that counted among its members the major unions representing pilots, flight attendants, and air traffic controllers. While there had not been any serious injury from a reported drone incident, Wytkind wrote, "there is little doubt that we should not open the skies to widespread commercial and recreational drone use until we understand the challenge and fully implement enforceable safety rules. Our message to the FAA: don't succumb to pressure from big corporations like Amazon, politicians and lobby groups until we can safely and effectively integrate drones into our National Airspace System."[33]

National Drone Registry

Seeking to create "a culture of accountability and responsibility" among hobbyists, the Department of Transporation announced plans to establish a national registry to track recreational users of drones in October 2015. The registry would provide a means, however tenuous, of tracing back a rogue drone to its owner and thereby subjecting that person to civil or criminal penalties. Under Secretary Anthony Foxx, the department aimed to stand up the registry under FAA management by the end of the year, working at deliberate speed. The avenue it chose was to issue an "interim final rule," a provision of the Administrative Procedure Act that allowed a federal agency, upon finding "good cause," to produce a regulation without following the usual notice-and-comment process.

Within ten days of the announcement, the FAA named the members of a task force it assembled to develop the registration process. The twenty-six individuals represented aviation and unmanned aircraft trade organizations, drone and camera suppliers, and major retailers, including Amazon, Walmart, and Best Buy. Earl Lawrence, who succeeded Jim Williams as director of the FAA's UAS Integration Office, and Google X Project Wing director Dave Vos were appointed to serve as co-chairmen.

The Unmanned Aircraft Systems Registration Task Force Aviation Rulemaking Committee met for three days in early November and delivered its recommendations to FAA Administrator Michael Huerta as directed on November 21. The committee called for creating a free, electronic registration system through which drone owners would enter their name and street address and receive a certificate of registration and a personal universal registration number they could use for multiple machines. Providing an email address, telephone number, and aircraft serial number would be optional and US citizenship and residence status would not be required. The owner would mark or affix the FAA-issued registration number on each drone so it could be identified. The committee believed this construct—electronic registration, assignment of a personal registration number, and requiring the number be marked on each drone—was the best approach under the rushed circumstances. "Any registration steps more burdensome than these three simple steps may jeopardize the likelihood of widespread adoption and would undermine the overall registration philosophy that enabled the Task Force to come to consensus," it advised.

After spending "considerable time" deliberating what the minimum weight threshold should be to trigger registration, the committee settled on 250 grams—about half a pound—with fifty-five pounds already accepted as the outer limit of what constituted a small unmanned aircraft system. In making that determination, it used an algebraic aviation risk assessment formula and the findings of a 2012 MITRE analysis of the kinetic energy of small moving objects, including drones, and their probability of causing a fatality on the ground. At the agreed registration threshold, there was a probability of one fatality for every twenty million flight hours, which "seems to present a reasonably acceptable risk level," the committee said."While general agreement was ultimately reached on the 250 gram weight, there were Task Force members who believed it was too conservative, as the weight could negatively impact the credibility of the sUAS registration program and thus lessen compliance levels because it would require registration of some sUAS generally considered to be in the 'toy' category," the group stated. "Task force members representing manned aircraft organizations expressed specific concerns that data on UAS-aircraft collisions, engine ingestion, propeller, and rotor impacts by UAS was not available when determining the weight threshold." The committee also decided another threshold: drone registration would apply to persons aged 13 or older. Anybody younger would operate under their parents' registration.[34]

Foxx and FAA Deputy Administrator Michael Whitaker revealed the "Registration and Marking Requirements for Small Unmanned Aircraft" interim final rulemaking to reporters in a morning teleconference on December 14, 2015. The rule would become effective on December 21, when it was published in the *Federal Register*. Although expedited, it was carefully constructed. On paper, the rule would span 211 pages of single-spaced text, but only ten pages actually described its requirements, former FAA general counsel J. E. "Sandy" Murdock III observed in a blog post. "It takes roughly 180 pages to introduce and explain the rule; included in that section are about thirty pages written to make a finding of 'good cause,'" Murdock wrote. "Less than five percent of the verbiage is the actual rule."

The FAA had taken the advice of the task force to establish a "streamlined and simple" registration system that it would host on its website. It also accepted the task force's recommendation of 250 grams (0.55 pounds), including ancillary camera or other attachments, as the baseline weight to register a drone, and thirteen as the minimum age requirement. However, the agency would charge a five dollar fee to register, which it offered to refund to drone owners who logged in within thirty days of the system's activation. A registrant's name, address, and email address would be required to enter the system—information the government could authenticate through a credit card transaction. The requirement was retroactive; those owners who operated drones prior to December 21 were given until February 19, 2016, to register. All drone owners would download a certificate of registration good for three years or a "certificate of ownership" in the case of foreign nationals and tourists. Failure to register brought the possibility of civil fines ranging up to $27,500 and criminal penalties of up to $250,000 or imprisonment. "The goal is not to be punitive, but to get people into compliance with the operating rules," Whitaker said. Within two days of the system's activation more than 45,000 people had registered, a small number of whom reported "performance issues" with the online site, the FAA said. With Christmas day looming and the prospect of hundreds of thousands of new drone owners joining the system, the agency instructed its contractor to take down the site for maintenance during overnight hours.

A mixed reception greeted the interim final rule. Trade organization AUVSI had participated on the rulemaking task force and declared itself pleased with the outcome.

"Task force members worked diligently to resolve their differences and find solutions that allow for the growth and development of the UAS industry while providing increased accountability," the association stated. "Though it may not be perfect, this process and final rule shows that industry and government can come together quickly to develop policy." The Academy of Model Aeronautics also participated on the task force and proclaimed itself disappointed. The rule violated the intent of Congress expressed in Section 336 of the FMRA legislation—the Special Rule for Model Aircraft—which prevented the FAA from regulating model aircraft that were flown within the safety guidelines of a "community-based organization." In June 2014, when the FAA issued an interpretation of the rule that expanded the definition of aircraft to include flying models, the AMA petitioned the US Court of Appeals for the District of Columbia to have the interpretation overturned. With that litigation still unresolved, the academy advised its membership to "hold off on registering" through the FAA system. "While we continue to believe that registration makes sense at some threshold, and for flyers operating outside of a community-based organization or flying for commercial purposes, we also strongly believe our members are not the problem and should not have to bear the burden of additional regulations," the AMA's executive council stated in a message posted on the academy's website.

Commercial Exemptions Abound

The FAA still required people who flew drones for business to register their aircraft through a paper-based system, so for a time there were two registration systems. It planned to transition to a web-based system for commercial operators in 2016. Under Section 333 of the FMRA legislation, the FAA started accepting petitions to operate drones commercially in May 2014. It granted the first exemptions to six Hollywood-affiliated film and production companies some four months later, on September 25. As more petitions arrived from aerial photography and surveying companies, real estate firms, farming interests, building contractors, utilities, and other industries, the agency relaxed the way it administered them. In March 2015, the FAA announced that operators who obtained exemptions would automatically receive a "blanket" certificate of authorization to fly their machines at or below 200 feet anywhere in the country, except in restricted airspace or in major cities. Other conditions still applied: the flights were restricted to daytime conditions, within the operator's visual line of sight, and at specified distances from airports or heliports, but no longer would the FAA set conditions for each operation flown within those parameters. In April 2015, it announced a "summary grant" process—a more expeditious method of approving batches of exemptions in cases that were similar to previous approvals. Two other changes were made to the exemption process. Whereas early Section 333 operators needed at least a private pilot's license to fly their machines, the FAA now allowed operations by people holding a recreational or a sport pilot certificate, which took less total flight time and money to earn. They no longer needed an FAA third-class medical certificate; a valid driver's license would satisfy the agency's medical requirement.

Twenty-two petitioners obtained exemptions from the FAA on April 8, 2015, among them Amazon. Nine months after applying, the on-line retailer was authorized to fly its Prime Air multi-rotor drone for outdoor testing over company-owned property with twenty-eight conditions attached. On May 1, the FAA granted Yamaha Motor

Corporation USA an exemption to operate the RMAX for "agricultural-related services." The authorization stipulated the helicopter's maximum takeoff weight should not exceed 218 pounds, at the time making it the largest unmanned aircraft platform the agency had approved for commercial flights. By August, the FAA had granted more than 1,000 Section 333 exemptions.[35] It was approving more than fifty applications per week.

Holding a Section 333 exemption did not excuse a company from operating without what the FAA determined was the requisite authority. SkyPan International, an aerial photography company based in Chicago, applied for an exemption to fly an Align T Rex 700E F3C model helicopter "to perform inspections of unimproved or vacant properties for the construction industry, developers and property owners" in December 2014. The FAA granted the exemption in April 2015, roughly a week after Amazon's approval. But within months the agency lowered the enforcement boom on SkyPan for its previous activities over Chicago and New York. In October, the FAA proposed a $1.9 million fine of the company for endangering airspace safety, its largest civil penalty to date. For three years prior to its Section 333 application, SkyPan "conducted 65 unauthorized operations in some of our most congested airspace and heavily populated cities, violating airspace regulations and various operating rules. These operations were illegal and not without risk," the agency charged in a press release. The highest previous civil penalty the FAA assessed was against Xizmo Media of New York, a self-described boutique production company that specialized in drone cinematography, for $18,700. Most other fines ranged between $1,100 and $5,500.

In a lengthy response posted on its website, SkyPan said it had conducted aerial photography over private property in urban areas for twenty-seven years in full compliance with published FAA regulations. It had "proactively contacted" the agency in 2005, 2008, and 2010 "to explore special permitting" for its planned commercial drone business. "SkyPan operates only in privately owned airspace over the private property of its clients," the company argued. "Never flying over people or public places, SkyPan always maintains straight up/straight down flights with full-time altimeter monitoring and complete aircraft control. SkyPan never operates higher than surrounding structures and never penetrates navigable US airspace as defined by the FAA." The company looked forward to a "comprehensive regulatory agreement" with the government, the statement added.

In the summer of 2015, AUVSI released an analysis of the first 500 commercial drone exemptions the FAA awarded. The approvals, some of which the agency granted for multiple purposes, spanned more than twenty industries overall. Real estate applications of drones were cited in 153 of the approved applications, followed by 128 mentioning aerial surveying, 125 aerial photography, and 106 agriculture. The operators came from forty-eight states, led by California with seventy approved operations, Texas with forty-six, and Florida with forty. When AUVSI updated the analysis in September to include the first 1,000 exemptions, aerial photography had moved into first place with 512 approvals, followed by real estate and aerial surveying.

The early preponderance of drone use or intended use by realtors, at least for the moment, belied the anticipated demand from farming—the segment AUVSI had predicted would lead all others in its March 2013 economic impact study. FAA conditions that required line-of-sight and daytime operations may have contributed to "inhibited adoption" by farmers early on, the association said. Precision farming also required

Drone hobbyist Larry Kelley of Amherst, Massachusetts, hand captures his DJI Phantom 2 quadcopter. An Association of Unmanned Vehicle Systems International analysis of the first 500 exemptions the FAA awarded for commercial uses of small unmanned aircraft revealed that DJI machines were dominant among early users. *Author photo*.

sophisticated, backend processing of aerial data; conversely, real estate professionals could upload the digital video and still images collected by relatively inexpensive drones to create web-based property listings. Indeed, consumer and "prosumer" grade quadcopters manufactured by China's DJI—the Phantom 2, Phantom 2 Vision +, and Inspire 1, ranging in price from around $600 to $3,000—dominated the field among early commercial users. DJI also manufactured the Spreading Wings series of hexacopters and supplied 446 out of 681 total platforms, or 65% of the models covered in the first FAA exemptions.

Google filed a Section 333 petition with the FAA in June 2015 for its Project Wing machine, which it described as a "Google-designed and manufactured multi-engine, battery-powered, vertical takeoff and landing, fixed-wing, small UAS." The late filing relative to other drone developers and operators did not mean the search engine giant was late to the game. At the time of the filing, Google already had a research relationship with NASA's Ames Research Center at Moffett Federal Airfield near Mountain View, California—the home of Google's headquarters in the heart of Silicon Valley. Under the Space Act Agreement, they would collaborate on "Autonomous Vehicle Operations Concepts and Technologies to Enable New Capabilities in Air Transportation." An online record of active "SAAs" revealed this as a five-year agreement dating from June 27, 2014, to June 27, 2019. Separately, NASA announced in November 2014 it had signed a long-term lease agreement with a Google subsidiary, Planetary Ventures, giving the latter entity management oversight of Moffett Field. Planetary Ventures planned to invest more than $200 million on improvements to the property, among them restoring Hangar One, a landmark structure the Navy built in 1932 to house the airship USS *Macon*.[36] In October 2015, Google underwent a corporate reorganization to create a holding company—Alphabet Inc.—and several independent subsidiaries, including the Google X research and development laboratory.

Already engaged with the FAA, General Atomics, Honeywell, and other partners in flying the five-ton Ikhana for collision avoidance research, NASA was taking a lead government role in the science around small drones. Around 2013, an Ames team led by Parimal Kopardekar, a NASA principal investigator and former FAA engineer who specialized in air traffic control automation, undertook the development of a system to manage drone traffic at low altitude—a construct that, if achieved, could make Amazon's and Google's far-fetched plans to deliver packages by drone work after all. The Unmanned Aerial System Traffic Management (UTM) system was established as a technical challenge under NASA's Safe Autonomous Systems Operations Project. Officially started in fiscal 2015 with $15.6 million in funding, the UTM effort "was extensively discussed internally and with potential partners for well over a year prior to that," the space agency said, in response to an inquiry. It received $600,000 in funding in fiscal 2014.

Kopardekar—"P.K." to those who knew him—described the UTM system as "a digital, or a virtual infrastructure" that would organize low-flying drone traffic by lanes and altitude limits in space. NASA planned to achieve the vision in four increments (software builds) supporting aircraft spacing, collision avoidance, and trajectory management functions for increasingly dense levels of drone traffic.

UTM1 would "create, analyze and manage (aircraft) trajectories and constraints that enable operations by an interactive system," the space agency said. "The focus will be on geo-fencing, altitude 'rules of the road,' and scheduling of vehicle trajectories." NASA planned a first-build demonstration in late August 2015 at Crows Landing Airport in central California to prove the concept. This would involve "a diverse set" of unmanned aircraft, all weighing less than fifty-five pounds. Through the successive builds, its goal was to create a prototype UTM system by 2019.

More than 120 mostly private companies, but also several universities and government agencies responded to a formal NASA request for information to collaborate on the UTM project. "Those organizations that responded to the RFI are offered opportunities to discuss the concept and provide feedback in various ways to help push the development of the concept and the prototype," NASA said. "Several of these partners have more formally engaged with us to perform joint testing of the system. We anticipate that the number of collaborators will grow as the effort advances."

NASA Ames hosted a "UTM 2015" convention at Moffett Field in July 2015. In a keynote address Gur Kimchi, cofounder of Amazon's Prime Air concept, laid out the retailer's vision of a future airspace system. It would contain segregated (restricted) airspace below 500 feet, leaving the zone closest to the ground for small drones and effectively separating them from manned aviation. The lower zone would be further stratified. The airspace below 200 feet would be reserved for localized traffic, such as for photography, surveying or inspection operations, and flights of less sophisticated drones. The airspace between 200 and 400 feet would be a "high-speed transit" layer for more sophisticated machines with sense and avoid capability flying beyond the operator's line-of-sight. Amazon even proposed a new model of airspace governance. "While more research is needed to identify exactly how an air navigation service provider will evolve to support high-volume sUAS operations, it is Amazon's position that the projected industry growth also requires the delegation of responsibility for many traditional air navigation services," the retailer declared in a white paper released at the convention. "[M]any of these services will be handled in a more distributed and federated fashion where multiple operators cover overlapping areas, each managing their own fleet."[37]

Besides Amazon and Google, other large companies harbored drone ambitions, seeing the potential of unmanned aircraft for deliveries and other applications. In March 2015, Motorola Solutions announced it had made a "strategic investment" in CyPhyWorks, a small company with patented microfilament tether technology that supported long duration flights and high-speed data streaming by tethered drones. Operating from a business park in Danvers, Massachusetts, CyPhyWorks was headed by MIT-trained roboticist Helen Greiner, who earlier co-founded iRobot, manufacturer of the "Roomba" robotic vacuum cleaner. The UPS Strategic Enterprise Fund, the private equity investment arm of delivery giant UPS, joined Motorola Solutions and other partners in a $22 million investment round to finance CyPhyWorks that October. "UPS invests more in technology than any other company in the delivery business, and we're always planning for the future," the company said in a statement provided to the author. "UPS is evaluating autonomous systems technology, including drones. We are currently experimenting with drone use in warehouse facilities, and participating in test scenarios supporting humanitarian relief efforts." In October 2015, retail multinational Walmart applied to

The popular Yuneec Q500 Typhoon quadcopter challenged DJI's dominance of the consumer drone market. In August 2015, Intel Capital, the investment arm of semiconductor chipmaker Intel, announced a $60 million investment in Hong Kong-based Yuneec. *Photo courtesy of Matt Thurber*

the FAA for a commercial exemption to test DJI multi-rotor drones for applications, including surveillance of its distribution center buildings and parking lots and ferrying groceries from retail outlets to customers in the parking lot. Semiconductor manufacturer Intel invested $60 million in Hong Kong-based drone manufacturer Yuneec Electric Aviation in August 2015; in January 2016, the chipmaker announced the acquisition of German drone maker Ascending Technologies. The drone manufacturers provided Intel with flying platforms to further develop its computer processors and depth-sensing "RealSense" camera for collision avoidance and other functions.

The unmanned aircraft industry was evolving. The center of gravity seemed to be shifting more toward the innovation of Silicon Valley and NASA and away from the bureaucracy of Washington, DC, and the FAA. At an Air Line Pilots Association conference in the nation's capital, Google X's Dave Vos spoke of a "really exciting and transformational future that's happening here," with the potential of billions of small packages being delivered by drone. "Silicon Valley is getting behind" the industry, he informed the audience. Long-time former Department of Defense executive Michael Toscano retired as CEO of trade organization AUVSI in January 2015. Brian Wynne, who previously headed the Electric Drive Transportation Association, succeeded him. That spring, AUVSI announced a new slate of directors that included its own former general counsel, Ben Gielow, by then serving as Amazon senior manager of public policy, and Dave Vos representing Google X. The association's Unmanned Systems conference was rebranded—it would be called "Xponential" to reflect the industry's "tremendous growth." Other conferences popped up to challenge AUVSI's primacy over the industry. The inaugural Drone World Expo held in San Jose in November 2015 attracted seventy-five exhibitors and 2,120 attendees—an "extremely successful launch" according to its organizers. The second annual International Drone Expo in Los Angeles the following month "was a huge success," with more than 100 exhibitors and 3,000 participants.

People who played a role in the industry's transformation moved on to other jobs, some still linked to UAVs. After his long career with the FAA, in 2014, Gary Michel joined the law firm Wiley Rein and its unmanned aircraft systems working group as a consulting counsel. In early 2015, Aerospace Industries Association chief executive Marion Blakey was appointed president and CEO of engine manufacturer Rolls-Royce North America. Blakey's former vice president of communications, Francis "Chip" Sheller, was named head of public affairs of UAV manufacturer Aurora Flight Sciences. After the leadership of the Air Line Pilots Association changed hands, Amazon hired its former second-ranking officer, Alaska Airlines pilot Sean Cassidy, as its director of strategic partnerships. Phantom manufacturer DJI hired attorney Brendan Schulman, renowned as "the drone lawyer" for his brilliant defense of Raphael Pirker, as vice president of policy and legal affairs. By then he had been immortalized in song, "The Ballad of Brendan Schulman," set to the music of Davy Crockett's anthem by self-professed Drone Singer Victor Villegas. Jim Williams, the first person to manage the FAA's UAS Integration Office, retired from the agency in June 2015. The next month, the law firm Dentons appointed him as a principal of its public policy and regulation practice. "Jim advises Dentons' clients on issues related to aircraft and space operations with specific emphasis on all things related to unmanned aircraft systems—a.k.a. drones," the firm said. Recognizing the proliferating demands of the job, the FAA appointed two men to replace Williams: Lawrence, who previously served as director of the agency's small airplane directorate; and retired Air Force two-star general Marke "Hoot" Gibson. The FAA named Gibson, a former command pilot and fighter wing commander, to the new position of senior advisor on UAS integration, reporting directly to the deputy administrator.

Export Control

Unmanned aircraft manufacturers and suppliers who sought to sell their products overseas had to comply with US export controls. Under an export reform initiative that entered force for aircraft, gas turbine engines, and related components in October 2013, the Obama administration shifted items considered non-sensitive or having dual military and commercial uses from the US Munitions List (USML) to the Commerce Control List (CCL). The State Department administered the USML; the Department of Commerce the CCL. The USML was subject to so-called ITAR (International Traffic in Arms Regulations) that controlled the export of military related items.

Moving certain items from the USML to the CCL allowed the United States to better focus its resources on items requiring the highest level of export scrutiny, while providing American companies with a means to sell less sensitive parts and components overseas, the administration explained.

The US aerospace industry had long advocated for such reform; it argued that strict export controls limited the growth of the nation's satellite industry in the 1990s, when satellite technology was subject to ITAR restrictions. "Export restrictions are hurting this industry without making us any safer, and they could cause the US to relinquish to other nations its lead in these technologies," Northrop Grumman CEO Wes Bush warned in a keynote speech to the 2011 Unmanned Systems conference. "Today, the US is struggling to sell unmanned aircraft to its allies while other nations prepare to jump into the marketplace with both feet. In a repeat of the satellite example, the thinking seems to be that our allies will neither build their own, nor buy them from those who will be motivated, by the perversity of our policies, to build them themselves."[38]

But export control reform in itself would not lower barriers to selling certain components and systems overseas said Eric McClafferty, an attorney specializing in international trade with Washington, DC, law firm Kelley Drye. "Except on the margins, export control reform is not going to have very much of an impact on the unmanned vehicle area, and the main reason for that is the existence of the Missile Technology Control Regime," he told the author. Founded in 1987 by the G-7 industrialized nations, the MTCR described itself as an informal, voluntary association of thirty-four nations committed to stopping the spread of "unmanned delivery systems capable of delivering weapons of mass destruction." Whereas US export control reform involved a unilateral movement of certain items from the State Department to the Commerce Department control lists, the restrictions imposed by the multi-lateral MTCR nevertheless remained the same, and the United States was not free to change those rules without consulting the other nations, McClafferty explained. He knew of preliminary efforts within Congress and the State Department to liberalize the MTCR to allow more commercial activity, but no major progress toward that end.

At the Farnborough International Airshow held outside London in July 2014, Ken Handelman, deputy assistant secretary of state for defense trade controls at the State Department, said the MTCR factored in an internal policy review of UAS exports the government was then conducting. "The Missile Technology Control Regime, which remains a hugely important tool for US policy and for the nonproliferation policy of all of our allies sometimes has unintended consequences so yes, that's been part of the consideration," Handelman said.

The MTCR listed as its main objective the control of goods and technologies that contributed to delivery systems for weapons of mass destruction, with a particular focus on rockets and UAVs capable of delivering a payload of at least 500 kg (1,102 pounds) to a range of at least 300 km (186 miles), "and on equipment, software and technology for such systems." The United States required exporters to obtain a license before shipping any UAV with a range of 300 kilometers or more to most destinations, according to McClafferty. MTCR rules and criteria were reflected in the ITAR and even flowed into Commerce Department regulations for dual-use items, he said.

Industry executives complained that MTCR restrictions held back their companies, but not those of non-member countries. "Absolutely it's impacted our business," Ryan Hartman, Insitu senior vice president of programs and later CEO told reporters in June 2014. "From an industry perspective we'd like to see an MTCR policy that enables us to be on par with the rest of the international community. If you look at the capabilities and technology that are available in the international market and you look at what restrictions we have through the MTCR and other things, it is a fact that we're falling behind in the international market as it relates to the technology we can bring to market."

Unmanned aircraft systems and components that were not ITAR-restricted remained subject to Department of Commerce Export Administration Regulations, which were themselves restrictive. Items transferred from the USML to the CCL under export control reform, which fell under the "600 series" of classification numbers, came with many of the same restrictions as existed under the ITAR, McClafferty said. Most of the 600 series items were eligible for a "strategic trade authorization" license exception, allowing license-free exports to thirty-six countries, but the STA exception did not cover items controlled through the MTCR. "There are a lot of folks out there who think that you can take a small unmanned vehicle and export it out of the US without any controls on it—something that can go less than 300 kilometers, which is the MTCR control level," McClafferty said. "The truth is, unless something is basically a toy or a model aircraft, if something can fly out of visual range, then you need an export license from the Commerce Department to send that out of the US, and that includes certain components of that type of an item as well."

Another effect of the reform was that many exporters would be interacting for the first time with the Commerce Department's Office of Export Enforcement—a unit that was more aggressive than the State Department in overseeing compliance with regulations, McClafferty said. Companies that discovered and voluntarily disclosed violations to the State Department would likely receive a warning letter; the Commerce unit was more inclined to issue a fine. In one case, the office announced in June 2014 that Fokker Services, a Netherlands-based company with a subsidiary in Atlanta, had agreed to pay a $10.5 million civil penalty for illegally exporting aircraft parts, technology, and services to Iran and Sudan, both of which were subject to US sanctions and Commerce licensing requirements.[39] "They have a bunch of enforcement agents and a budget for an enforcement office and managers and multiple field offices where they have people out conducting investigations," McClafferty related. "A lot of folks who have exclusively been at State under the ITAR are in for a wake-up call when they start to submit voluntary disclosures to the Commerce Department."

Unmanned aircraft manufacturers and suppliers were advised to do their homework when contemplating overseas exports, including knowing the control categories and how the items they manufactured were classified. "In many cases you can do a lot of what you want to do if you dig into the regulations and understand them and make your license applications," McClafferty said. "Once you learn how to complete the license applications in a way that Commerce wants to see them completed, then you can pretty much be assured that if you have a legitimate end user, you're going to get your license and go about your business."

In February 2015, the State Department announced a new export policy allowing sales of US-made military and commercial unmanned aircraft to other friendly nations on a case-by-case basis, a broader accommodation than it previously allowed. But an earlier, lesser-known policy change by an international grouping promised to further liberalize the sale of small, commercial drones outside of the United States, McClafferty told the Unmanned Systems 2015 conference. At the annual meeting of the forty-one-nation Wassenaar Arrangement in Vienna the previous December, participating nations agreed to new export controls in various categories, including UAVs. Under the revised controls, commercial UAVs with a flight endurance of less than thirty minutes would no longer require an export license from the Commerce Department, according to McClafferty's presentation. Commercial UAVs with a flight endurance between thirty minutes and an hour would only be subject to the license requirement if they could fly in wind gusts of twenty-five knots or more. "A new approach and language are coming for the export of commercial UAV systems," McClafferty said.

CHAPTER 11

UAVs ABROAD

Along the coast of Wales, where the land ascends from the seaside village of Aberporth, Ray Mann strived to make a go of the National Aeronautical Center. Its name was grand, but in the summer of 2014, the center was more a glimmer of Mann's vision of an important national asset. "If you build it, they will come"—a misquotation of the signature line from *Field of Dreams*—came to mind when contemplating the scene, a small aerodrome nestled in the green-hued quilt of the Welsh countryside. This was West Wales Airport, the foundation on which Mann dreamed of building an unmanned aircraft system development, test, and demonstration center. An entrepreneur and the owner of an electronics recycling business, Mann purchased the inactive former World War II airfield in 2001 to fly his plane back and forth from his country cottage. General aviation traffic was light, so he considered other ways to make the airport profitable. The September 11 terrorist attacks in the United States happened soon after he bought the airport; the military campaigns that followed brought to the fore unmanned aircraft.

Until they had an acceptable means to detect and avoid other aircraft, Britain's Civil Aviation Authority (CAA) restricted unmanned aircraft to segregated airspace, including airspace designated as "danger areas" because of military testing activities. West Wales Airport, a CAA-licensed civilian airport open to all pilots, lay within the "D201" danger area, covering some 2,000 square miles of Cardigan Bay to the west. The UK Ministry of Defense's Aberporth Range managed the airspace, which it used to test unmanned aircraft and missiles. The airport started hosting UAS demonstrations in 2004. That year, the Welsh Development Agency confirmed plans to develop a technology park on former MOD land adjacent to the airport—Parc Aberporth—as an unmanned aircraft center of excellence, promising to eventually create 1,000 jobs.[1] In September 2005, the company Thales UK conducted the first flight in the United Kingdom of a tactical unmanned aircraft—the Hermes 450 manufactured by Israel's Elbit Systems—on which the British Army based its Watchkeeper program. Mann applied for and received approval to host UAS flights in 2006 using a CAA-accredited UAS operations manual. At the request of the airport and the Welsh government, the CAA authorized another 499 square miles of segregated airspace over land to the east in June 2011, specifically for testing unmanned aircraft. The new airspace, designated D202, was only activated during UAS operations. The MOD's air traffic control unit at Aberporth provided a danger area crossing service, allowing manned aircraft to also use the airspace.[2]

Anti-war groups opposed to drones and residents of the area concerned over safety and noise challenged the Welsh government's ambition of building an economic cluster around unmanned aircraft from its inception. The Campaign for Nuclear Disarmament Cymru "was inspired to speak out against drones when they discovered in 2004 that the Aberporth training area in Wales—an area that is also a missile base—was slated to become a 'UAV Centre of Excellence,'" Medea Benjamin wrote in *Drone Warfare*. CND Cymru national secretary Jill Gough told Benjamin: "Quite apart from the problem that these machines and their imaging equipment were being tested over our homes, many people objected to the terrible fact that our community, and our country, was planning something appalling against people in other countries. We certainly don't want Wales to be part of that."[3] A West Wales-focused group called Bro Emlyn for Peace and Justice organized demonstrations and regularly shadowed the activities at the airport on its website. Parc Aberporth suffered a hit when anchor tenant Selex Galileo, part of Italy's Finmeccanica industrial group, announced in 2011 it was cutting operations there and elsewhere in the UK due to reduced defense spending. Selex had tested its Falco UAS at Aberporth, one of which crash-landed in September 2009, Bro Emlyn and local media reported.[4] The company later tested the Falco in Bulgaria.

Mann unveiled the National Aeronautical Center at the biennial Farnborough Airshow outside of London in 2012. West Wales Airport was a special piece of the nation's aviation infrastructure, he argued, because it offered companies the opportunity to test fly large UAS beyond a pilot's visual line of sight in regulated airspace that accommodated manned aircraft—a unique combination of factors. At an event held at the Royal Society in London in September 2013, Mann announced a partnership with Newquay Cornwall Airport in southwest England, another facility with aspirations but not the same expertise, for hosting unmanned aircraft operations. Whereas West Wales Airport offered a 3,900-foot runway, Newquay Cornwall, the former Royal Air Force St. Mawgan air base, boasted a 9,800-foot runway and access to 3,088 square miles of segregated airspace.

On the eve of the 2014 Farnborough Airshow in July, Mann awaited the second half of the *Field of Dreams* promise. With its two-story tower, West Wales Airport managed some 1,500 takeoffs and landings per year—not quite the bustling flight-test center he imagined. A few cars occupied spaces at nearby Parc Aberporth, but the technology park was wind-blown and tranquil. A knock at the offices of the West Wales UAV Center managed by defense contractor QinetiQ went unanswered. The airport did have a very significant customer—three months earlier, the MOD awarded it a two-year, £2.5 million contract to continue supplying its airfield and services for Watchkeeper test and evaluation. Watchkeeper airframes were trucked to Aberporth two at a time from manufacturer UAV Tactical Systems, the Thales UK and Elbit joint venture based in Leicester. Contractors assembled the cigar-shaped airplanes at a secure MOD facility adjoining the airport and ran them through their first ground and flight tests. The British Army had a requirement for fifty-four air vehicles, and on this day Watchkeeper number thirty-three made its first flight. Once the MOD tested and cleared the airplanes for use, they were repackaged in special containers and transported to the ministry's Boscombe Down test site or wherever required.

Unlike in the United States, Europe and other parts of the world were opening their skies to small unmanned aircraft through more flexible regulations than the FAA was prepared to deliver. But there were few programs to develop large, civil UAS, which

A British Army Watchkeeper UAV takes flight from West Wales Airport in Aberporth, UK, where contractors conducted shakedown flights. The Watchkeeper program was the anchor tenant of the National Aeronautical Center, which was having a difficult time attracting commercial customers. *Author photo*.

manufacturers would have to build to manned aircraft standards—a stumbling block all nations faced. With the same technological challenges of sense-and-avoid and command and control to overcome, other countries were years away from granting routine access to their airspace for larger machines.

The amiable Mann believed manufacturers lacked the will to advance the technology. "There doesn't seem to be a vision, there doesn't seem to be an appetite for them to spend their money to create something to go to market with," he lamented. "Most aviation companies are wedded to passenger airliners because they know that market, they're comfortable with that market, and others are wedded to the military budget because they know that market too and it's not their money they're spending. They have a wish, but they don't have a will." At an "RPAS Today" conference the Royal Aeronautical Society held in London around the same time, Mark Swan, director of the CAA's safety and airspace regulation group, said airspace the authority created for unmanned aircraft remained largely unused. "The Watchkeeper is in the system and cleared to fly, but there are no civilian systems (weighing more than 150 kg) operating in the UK," Swan said, according to *Aviation International News*. "I fully expected to have something big and beautiful flying in that airspace we set up by 2011, and it hasn't happened."[5]

In October 2014, the military division of the Airbus group made a step toward that vision by applying to the European Aviation Safety Agency (EASA) for civil type certification of its 570-kilogram Atlante unmanned airplane, which it was testing in Spain. This being the continent's first such application, Airbus said it would work with EASA to develop a certification process for the Atlante "based on the process normally used for manned aircraft, which will set the standards for future UAV certifications in Europe." French electronics concern Sagem completed a series of test flights of its Patroller UAV near Toulouse, France, in November 2014 that demonstrated "a complete anti-collision function." Sagem achieved sense-and-avoid capability using a combination of sensors, including an infrared optronic device and "automatic risk collision estimation and avoidance flightpath generation" software. The project proved the ability of the one-ton Patroller to conduct approaches to Toulouse-Blagnac airport based on air traffic control procedures.[6] None too soon NATS, the air navigation service provider of the United Kingdom, and manufacturer Thales said they performed the first flight of a large unmanned aircraft in non-segregated, controlled airspace over the island on September 30, 2015. The 1,050-pound Watchkeeper flew from West Wales Airport.

Fledgling flight test facilities waited in the wings. In early 2014, Mann's National Aeronautical Center; the CESA UAV test and service center near Bordeaux, France; the Oklahoma State University site in Elgin, Oklahoma; and the UAV Center of Excellence in Alma, Canada, formed the International Consortium of Aeronautical Test Sites. Two Spanish test facilities later joined the grouping: the Air Traffic Laboratory for Advanced Systems of Villacarrillo and the CATUAV Technology Center of Barcelona.

An alphabet soup of collaborations that industry and government formed to solve the technological challenges of flying drones plugged along. The European defense and space agencies funded the Demonstration of Satellites, enabling the Insertion of RPAS in Europe (DeSIRE) project. In April 2013, the partners flew an Israel Aerospace Industries' Heron 1 from San Javier Air Base in Murcia, Spain, in both civil and military airspace to demonstrate beyond line-of-sight control by satellite. In October 2013, the Brussels-based organization managing the Single European Sky Air Traffic Management Research effort (SESAR) selected nine demonstration projects to explore the feasibility of introducing remotely piloted aircraft in non-segregated airspace by 2016. The projects involved industry and government partners from the Czech Republic, France, Germany, Italy, Malta, the Netherlands, Spain, and the United Kingdom. Among them, the Operational Demonstration of RPAS in the European Airspace project that demonstrated the Patroller brought together Sagem, Rockwell Collins, France, the French Civil Aviation University in Toulouse, and France's air navigation service provider DSNA.

In the United Kingdom, the Autonomous Systems Technology Related Airborne Evaluation and Assessment program entered a third phase in 2014, aimed at gathering unmanned aircraft performance data through "extensive flight trials" to support regulatory decisions. The industry partners—software firm AOS, BAE Systems, Cobham, Cassidian, QinetiQ, Rolls-Royce, and Thales—over eight years had invested more than £30 million in the project; the UK government's Technology Strategy Board provided additional matching funds.[7] Under the aegis of the European Defense Agency, France, Germany, Italy, Spain, and Sweden launched the €50 million Midair Collision Avoidance System program in 2009, to develop an unmanned collision avoidance capability. During a MIDCAS-sponsored demonstration in the spring of 2015, Italian contractor Alenia Aermacchi successfully flew its Sky-Y demonstrator fitted with a sense-and-avoid suite against a C-27J military cargo aircraft flown as an intruder.

European and other countries aspired to build and sell indigenous UAVs for mainly military reasons, mounting a challenge to the longstanding dominance of the United States and Israel as suppliers of large military machines. In June 2009, Austrian manufacturer Schiebel became the first company to fly an unmanned aircraft—the Camcopter S-100—in the official flight display of the Paris Airshow, another major biennial arms bazaar. Turkey originally acquired Gnat 750 drones from General Atomics in the early 1990s; it first flew the indigenous Anka tactical UAV from Sivrihisar Air Base on December 30, 2010.[8] The Dassault Aviation-led Neuron unmanned combat air vehicle demonstrator first flew on December 1, 2012, from Istres, France.[9] Britain revealed the maiden flight of its secretive Taranis UCAV demonstrator in February 2014, six months after the fact. It was thought the flight took place in Woomera, South Australia, site of a Royal Australian Air Force test range.[10] At the Farnborough Airshow in July 2014, British and French governments agreed to jointly study a Future Combat Air System combining the Taranis and Neuron

The Camcopter S-100 helicopter, manufactured by Austrian company Schiebel, became the first unmanned aircraft system to perform in the official flight display of the Paris Air Show in June 2009. The 243-pound empty weight helicopter delivered real-time video and could fly for six hours. *Author photo.*

developments. Russia's famed aircraft manufacturers MiG and Sukhoi developed an unmanned combat UAV based on the MiG Skat demonstrator revealed in 2007. The Russian Sokol plant in Kazan built a prototype twin-turboprop UAV called Altius. India produced the Rustom series of tactical UAVs; Pakistan made the Buraq and Shahpar. Adcom Systems of United Arab Emirates built the serpentine, multi-winged Yabhon United 40 and other unmanned aircraft.[11]

China first flew its Lijian ("Sharp Sword") unmanned combat UAV demonstrator in November 2013. It also produced the Predator-like Wing Loong and the BZK-005 Giant Eagle, a large twin-boom aircraft powered by pusher propeller. China was also emergent as a manufacturer and exporter of consumer machines. Da-Jiang Innovations (DJI)—a company based in the special economic zone of Shenzhen, abutting Hong Kong—was becoming a major player in small drones. The privately-owned company started business in 2006. It declined to provide sales figures, but in response to an inquiry said its revenue had grown by three to five times year over year between 2009 and 2014, a period that saw it grow from fifty to 2,800 employees. Best known as the name behind the popular Phantom quadcopter, DJI unveiled a sophisticated new model in 2014. The professional-grade Inspire 1 was also a quadcopter, but one with articulated rotor arms that lifted in flight to provide its rotating camera with an unobstructed, 360-degree view of the ground. Aimed at the aerial filmmaking market, the Inspire 1 sold for around

Chinese company DJI unveiled the professional-grade Inspire 1 quadcopter in 2014. It featured articulated rotor arms that lifted in flight to provide its rotating camera with an unobstructed, 360-degree view of the ground. Aimed at the aerial filmmaking market, the Inspire 1 retailed for around $3,000. *Courtesy of DJI*

$3,000 online. "2014 was the year that drones went from creepy military gear to consumer-friendly video cool, and it's almost entirely due to DJI," said the business publication *Fast Company*, which named DJI one of the world's fifty most innovative companies in 2015.

Small Unmanned Aircraft

Under European regulations EASA, based in Cologne, Germany, regulated RPAS having a maximum takeoff mass above 150 kilograms (330 pounds); civil aviation authorities in individual countries were responsible for aircraft weighing less. Thanks to national aviation authorities' relatively liberal—some might say enlightened—accommodation of small RPAS ranging from two to twenty or more kilograms, the lower end of the civil/commercial market for unmanned aircraft was erupting outside of the United States. In general, operators were restricted to flying their machines within visual line of sight, no farther than 1,640 feet from the pilot, no higher than 400 feet off the ground, and safely removed from an airport or populous area. Norway, Denmark, Sweden, Finland, Latvia, Lithuania, Poland, Germany, Switzerland, Czech Republic, Malta, Italy, Spain, France, Belgium, the United Kingdom, and Ireland had rules in place or under consideration for commercial and non-military governmental use of small unmanned aircraft. Peter van Blyenburgh, founder and impresario of Paris-based UVS International, estimated there were 1,700 certificated RPAS operators across Europe in 2014.

A Dutch national residing in Paris, van Blyenburgh was a large and boisterous figure with a crown of white hair, a deep voice, and a good command of American curse words. He became interested in unmanned aircraft while participating in the development of the Altec MART—translated as "mini aircraft for terrestrial reconnaissance"—which the French army fielded in Iraq during the first Gulf War in 1990–1991. In 1995, van Blyenburgh founded "Euro UVS," which a mission statement described as "an initiative to federate the European UVS industry." The organization was incorporated in Paris as a French-registered association in 1998. When French law governing associations proved restrictive, it shifted its legal status to the Netherlands in 2000. As the named implied, Euro UVS was initially focused on Europe and in its early years mainly on military machines. Within a few years, the association's general assembly voted to open its board and membership to non-European countries, and in 2004, it became UVS International. The nonprofit concern organized educational events around the world, including an annual "RPAS policy forum" in Brussels at which marketing pitches were strictly forbidden. Anyone who started hawking their product line was shown the door, van Blyenburgh said. "We became aware that this was going to spread out into the non-military market, and that this would be a big change not only for the industry but also for society. To make this possible, it was necessary to have rules and regulations," he said of the association's evolution. "Practically everything that is now ongoing outside the United States in the field of rules and regulations, we were either involved or the instigator. We were the ones that got the European Commission interested. We were able to prove that RPAS could create jobs and give society economic benefits." Asked if the effort was a "hard sell," van Blyenburgh replied: "It wasn't a hard sell; it was a long sell. You were dealing with a bunch of guys on three levels: the politicians; the regulatory authorities on a national level and on a European level; and the manufacturers and operators. Getting all these guys lined up to see the light in the same direction—that wasn't easy."

In April 2012, France's civil aviation authority (DGAC) published regulations that for the first time addressed RPAS involved in performing "aerial work." As of 2014, the agency had issued more than 500 authorizations representing more than 800 RPAS.[12] The guiding regulatory document in the United Kingdom—Civil Aviation Publication 722 (CAP 722)—required that anyone seeking to fly an unmanned aircraft weighing less than forty-five pounds for aerial work apply for and obtain CAA authorization. The CAA issued 318 permissions in 2013 and was on pace to exceed that number the following year.[13]

Elsewhere in the world, countries were seeing robust growth in the commercial use of drones. Canada's regulatory agency Transport Canada required operators to comply with the conditions of a Special Flight Operations Certificate, which it granted to applicants on a case-by-case basis for commercial and scientific missions. The agency issued 155 SFOCs for unmanned aircraft in 2011. It issued double that—347 SFOCs—in 2012, and then 945 SFOCs in 2013. Kevin Lauscher, director of industrial sales with multi-rotor helicopter manufacturer Draganfly Innovations, said his company applied for an SFOC in the spring of 2013 for a flight the company planned to make that summer in northern Canada under a special circumstance—it would be within 400 meters of an airstrip. "We had normally been getting authorization within two to three weeks," Lauscher related. "When we made the application we were told that the number of applications had increased dramatically from the previous year and to expect a four-month wait time." In November 2014, Transport Canada announced it would no longer require SFOCs for certain operations involving small unmanned aircraft weighing less than 4.5 pounds or between 4.5 pounds and 55 pounds. It required operators to comply with numerous safety conditions, among them keeping their models within the pilot's visual line of sight, below 300 feet, and at least five nautical miles from an airport. The authority prohibited first person view flights.

Australia's small drone industry was "booming," a parliamentary committee declared in July 2014. The "Eyes in the sky" report of the House of Representatives Standing Committee on Social Policy and Legal Affairs addressed both air safety and privacy implications of remotely piloted aircraft (RPAs) based on a series of hearings and roundtable discussions held earlier that year involving aviation authorities, academics, and industry groups.[14] The committee found that improvements in piloting and control technologies and reduced prices were driving sales of RPAs to hobbyists and aspiring business operators. "The Committee has heard that this has led to a large increase in the number of untrained RPA operators, many of whom are either unaware of or do not follow" aviation regulations, the report stated. "This presents a substantial risk to air safety." The committee noted an incident months earlier in which a small unmanned helicopter used to film the Endure Batavia Triathlon in Geraldton, Western Australia, allegedly struck and injured a woman participating in the competition. The operator denied the errant vehicle actually hit the woman and claimed it had been "hijacked" by someone nearby who was "channel hopping" radio frequencies.

As in other countries, Australian civil aviation safety regulations Part 101 distinguished between commercial and recreational use of RPAs and required operators seeking to fly unmanned aircraft for compensation obtain certification from the Civil

Aviation Safety Authority (CASA). The agency issued the first such certificate to aerial imaging and surveying company HELImetrex of Stafford Heights, a Brisbane suburb, in November 2002, and through January 2013 had issued twenty-seven certificates, according to the Australian Certified UAV Operators Association. As of the summer of 2014, CASA listed 109 certified operators and another forty applications were being processed, the agency told the parliamentary committee. By the fall, it had certified 150 operators. An analysis by the operators' association revealed that 70% of certificate holders were micro businesses employing less than five people. Small, multi-rotor helicopters formed the bulk of the nation's drone fleet, with the mining and agricultural industries providing early demand for services. With the exception of two Australian subsidiaries of American firms—AAI Aerosonde and Insitu Pacific—medium and large aerospace and defense companies were not "proactively seeking" RPA operator certificates from CASA.[15]

As the number of operator certifications climbed, CASA proposed updating and amending Part 101, by then twelve years old. One key provision would excuse operators of RPAs weighing 4.4 pounds or less from having to obtain either a remote pilot or an unmanned aircraft system operator's certificate. The exception would be allowed, provided they followed standard RPA operating conditions: the aircraft would have to be flown within visual line of sight; less than 400 feet above ground or water; more than 100 feet from any person not involved in the operation; during daytime only; outside of controlled airspace; and more than three nautical miles from an airport. "CASA has investigated the risks and the potential for harm to people and property on the ground and other airspace users associated with impacts from small RPA in order to determine a low kinetic energy RPA mass," the agency's notice of proposed rulemaking stated. "The general consensus is that RPA with a gross weight of 2 kilograms (4.4 pounds) and below have a very low kinetic energy, pose very little risk to aviation and have a low potential for harm to people and property on the ground and other airspace users."[16] In its report, the parliamentary committee deemed CASA's amended rule to be "appropriate" and striking a suitable balance between minimizing safety risks and facilitating the growth of Australia's RPA industry. The operators' association disagreed. "ACUO…is deeply concerned that the Committee has stepped outside its terms of reference to provide unqualified support for the proposed deregulation of the sub-2kg category of remotely piloted aircraft systems," the association said. "[I]t is the sub-2kg class of systems, operated by hobbyists, which are already the primary source of near miss incidents involving manned aviation, and of emerging privacy complaints."[17]

In van Blyenburgh's estimation, the commercial RPAS industry was "very much helped" by regulators in Europe and elsewhere. "It's beyond me when I hear the press talk about what's going on in the United States," he said, by way of comparison. "Flying a ScanEagle to look for seals in Alaska is not a commercial operation. But this is being touted by the FAA and God knows, the unknowing press, as being a great feat. It is not." What Europe lacked was uniform regulations. "The down side is we have 28 countries in the European Union. All of these national initiatives, even though they can be applauded, are all non-harmonized, which means that an approved operator in France can still not fly in Sweden," he added.

In June 2013, the European RPAS Steering Group on which van Blyenburgh served, a collaboration of air traffic management, safety, and standards organizations assembled by the European Commission, released a "roadmap" to progressively introduce unmanned aircraft into civilian airspace beginning in 2016. The authors recognized that fifteen countries had already developed regulations and others were in the process "due to the mounting pressure from civil RPAS operators." What countries lacked was mutual recognition of regulations across borders. A group of experts called the Joint Authorities for Rulemaking on Unmanned Systems, representing national civil aviation authorities from European and other countries, was at work "harmonizing" or aligning such regulations. Once that was accomplished, the roadmap authors proposed transferring responsibility for RPAS weighing less than 150 kilograms from the national civil aviation authorities to EASA, which already regulated heavier vehicles. They recommended research and development initiatives supporting the introduction of RPAS be addressed under the umbrella of the SESAR program, the technology pillar of the continent's vision of a "Single European Sky." Future RPAS regulations were to incorporate "a clear liability regime and an adequate insurance obligation" to compensate for injuries or damage suffered by other airspace users or third parties on the ground. They should address the security risks of RPAS being hijacked or used as weapons and comply with privacy and data protection regulations. In the roadmap's end state period of 2024 to 2028, RPAS "are expected to operate in most non-segregated airspace, mixed with manned aviation" at an equivalent level of safety and security. "[B]ased on common rules, acquired experience and mutual recognition established among member states, RPAS operators could fly cross border intra EU, based on the 'file and fly' principle, and avoid the administrative burden to apply for special authorization before filing the flight plan."[18]

Within a year after the roadmap's release, the European Commission signaled its intention to set "tough new standards" for the safety, security, privacy, data protection, and insurance of what it called civil drones. It promised "industrial support" of start-ups and small and medium-sized enterprises involved in the sector through the European Union's Horizon 2020 research and "COSME" competitiveness programs. "The aim is to allow European industry to become a global leader in the market for this emerging technology, while at the same time ensuring that all the necessary safeguards are in place," the commission said. "Civil drones are increasingly being used in Europe...but under a fragmented regulatory framework. Basic national safety rules apply, but the rules differ across the EU and a number of key safeguards are not addressed in a coherent way."[19] In a statement accompanying the announcement, Siim Kallas, the Estonian politician who then served as European commissioner for transport, said: "[M]any people, including myself, have concerns about the safety, security and privacy issues relating to these devices...If ever there was a right time to do this, and to do this at a European level, it is now."[20]

On December 7, 2015, the European Commission produced a new "strategy" document that aimed to revise aviation safety policy dating to 2008. Among the commission's goals was to "unleash the full potential of drones" by creating a legal framework for their commercial operation that would apply across the continent. On December 18, EASA published a "technical opinion" that described that framework. The proposed construct established "open," "specific," and "certified" categories for drone operations based on their associated risk, with minimal requirements for the open category ascending to stricter requirements comparable to manned aviation for the

Torquing Group, of Pembrokeshire, Wales, displayed the AV Sparrow at the Unmanned Systems 2014 conference in Orlando. The miniature quadcopter weighed 3.5 ounces and used sonar to sense and avoid indoor objects. *Author photo.*

highest risk certified category. "Proportionality is a key, if not the most important feature of the regulatory framework," the agency said.

The International Civil Aviation Organization, which developed standards and recommended practices its 191 member states followed worldwide, turned up its own efforts on drones. It upgraded its unmanned aircraft systems study group formed in 2007 to a full panel, which ranked higher in the organization's hierarchy. At its first meeting in November 2014, the RPAS panel elected as its chairman Randy Willis, an executive with the FAA's UAS Integration Office. Mike Gadd, manager of aircraft certification with the British CAA, was named vice chairman. The panel aimed to deliver standards for airworthiness, operations, and pilot licensing to ICAO's governing council in 2018.

ENDNOTES

Chapter 1

1. Tad McGeer, "Laima: The First Atlantic Crossing by Unmanned Aircraft," The Insitu Group, February 25, 1999.
2. Jim Garamone, "ScanEagle Proves Worth in Fallujah Fight," American Forces Press Service, January 11, 2005, http://www.defense.gov/news/newsarticle.aspx?id=24397.
3. The Boeing Company, "Boeing Acquires Insitu to Expand Capabilities in Unmanned Systems," news release, July 22, 2008, http://boeing.mediaroom.com/2008-07-22-Boeing-Acquires-Insitu-to-Expand-Capabilities-in-Unmanned-Systems.
4. VT Group, "Shell Oil in the Arctic, UAV Operations in Commercial Environments," case study, http://www.vt-group.com/case-studies/shell-oil-in-the-arctic/.
5. Royal Dutch Shell, "Shell in the Arctic," briefing note, April 2011.
6. Rob Stapleton, "Shell plans unmanned flights over oil seas," *Alaska Journal of Commerce*, August 3, 2008, http://classic.alaskajournal.com/stories/080308/hom_20080803002.shtml.
7. Dale W. Funk, et al., "Joint Monitoring Program in the Chukchi and Beaufort seas, open water seasons, 2006–2008," LGL Alaska Research Associates, Greeneridge Sciences and Jasco Research for Shell Offshore, March 2010, http://s00.static-shell.com/content/dam/shell/static/usa/downloads/alaska/2006-2008-comprehensivereport.pdf.
8. Bill Carey, "FAA Plans Unmanned 'Sense and Avoid' Rule in 2016," *AINonline*, July 22, 2013, http://www.AINonline.com/aviation-news/ain-air-transport-perspective/2013-07-22/faa-plans-unmanned-sense-and-avoid-rule-2016.
9. Andrew R. Mahoney, "Sea Ice Conditions in the Chukchi and Beaufort Seas," Pew Charitable Trusts, November 2012.
10. US Bureau of Ocean Energy Management, "Assessment of Undiscovered Technically Recoverable Oil and Gas Resources of the Nation's Outer Continental Shelf, 2011," November 2011, Washington, DC, http://www.boem.gov/uploadedfiles/2011_national_assessment_factsheet.pdf.
11. ConocoPhillips, "Plan of Cooperation, Chukchi Sea Exploration Program, Devils Paw Prospect," February 2012, Anchorage, Alaska.
12. The Native village of Point Hope, et al., "Opposition Builds Against Drilling Arctic's Chukchi Sea," news release, January 31, 2008, http://northern.org/media-library/document-archive/arctic/arctic-ocean-reports-fact-sheets/alaska-natives-

and-conservation-groups-challenge-chukchi-sea-lease-sale-193/at_download/file.
13. Dan Joling, "Drifting sea ice halts Shell's Arctic drilling," The Associated Press in *Alaska Dispatch News*, September 10, 2012, http://www.adn.com/article/20120910/drifting-sea-ice-halts-shells-arctic-drilling.
14. Lisa Demer, "Kulluk grounding: Coast Guard report finds series of failures, questionable decisions by Shell," *Alaska Dispatch News*, April 3, 2014, http://www.adn.com/article/20140403/kulluk-grounding-coast-guard-report-finds-series-failures-questionable-decisions.
15. Alan Bailey, "Patrolling the Arctic," *Petroleum News*, November 5, 2006, http://www.petroleumnews.com/pntruncate/706842467.shtml.
16. ConocoPhillips, "Plan of Cooperation, Chukchi Sea Exploration Program, Devils Paw Prospect," February 2012, Anchorage, Alaska.
17. Patrick Egan, "ConocoPhillips confirms ScanEagle down," *sUAS News*, October 2, 2013, http://www.suasnews.com/2013/10/25402/conocophillips-confirms-scaneagle-down/.
18. AeroVironment, "BP and AeroVironment Launch First FAA-Approved Commercial Unmanned Aircraft Operations Over Land and Water; Providing Comprehensive GIS Services," news release, June 10, 2014, http://www.avinc.com/resources/press_release/bp-and-aerovironment-launch-first-faa-approved-commercial-unmanned-aircraft.
19. The six aerial photo and video production companies awarded Section 333 exemptions by the FAA were Astraeus Aerial of Los Angeles; Aerial MOB of Los Angeles and San Diego; Pictorvision of Van Nuys, California; HeliVideo Productions of Austin, Texas; Snaproll Media of Franklin, Tennessee; and RC Pro Productions Consulting/Vortex Aerial of Huntington Beach, California.

Chapter 2

1. Philip Hammond, "In defence of drones," *The Guardian*, December 17, 2013, http://www.theguardian.com/commentisfree/2013/dec/18/in-defence-of-drones-keep-civilians-troops-safe.
2. FlightGlobal/Archive, "Automatic Flight, The Forty-Sixth Wilbur Wright Memorial Lecture," May 16, 1958, http://www.flightglobal.com/pdfarchive/view/1958/1958%20-%200642.html?search=Automatic%20Flight.
3. John F. Keane and Stephen S. Carr, "A Brief History of Early Unmanned Aircraft," Johns Hopkins *APL Technical Digest*, Vol. 32, No. 3, 2013.
4. Linwood S. Howeth, *History of Communications—Electronics in the United States Navy*, Bureau of Ships and Office of Naval History, Washington, DC, 1963, p. 481.
5. Steven J. Zaloga, *Unmanned Aerial Vehicles, Robotic Air Warfare 1917–2007*, Osprey Publishing, 2008, p. 8.
6. National Museum of the US Air Force, "Radioplane/Northrop MQM-57 Falconer," fact sheet, October 22, 2013, http://www.nationalmuseum.af.mil/factsheets/factsheet.asp?id=7684, and Zaloga, p. 10.
7. Zaloga, p. 16.
8. Public Law 107-40, "Authorization for Use of Military Force," September 18, 2001, US Government Printing Office, Washington, DC, http://www.gpo.gov/fdsys/pkg/PLAW-107publ40/pdf/PLAW-107publ40.pdf.

9. Bob Woodward, *Bush at War*, Simon & Schuster, New York, 2002, p. 101.
10. Jonathan Masters, "Backgrounder: Targeted Killings," Council on Foreign Relations, May 23, 2013.
11. James Risen and Judith Miller, "CIA Is Reported to Kill a Leader of Qaeda in Yemen," *The New York Times*, November 5, 2002.
12. Seymour Hersh, "King's ransom," *The New Yorker*, October 22, 2001.
13. Richard Whittle, *Predator, The Secret Origins of the Drone Revolution*, Henry Holt and Company, New York, 2014, advance reader's edition, pp. 259 and 309.
14. Peter Bergen and Katherine Tiedemann, "Revenge of the Drones, An Analysis of Drone Strikes in Pakistan," October 19, 2009, and "The Year of the Drone, An Analysis of U.S. Drone Strikes in Pakistan, 2004–2010," February 24, 2010, New America Foundation, Washington, DC.
15. Human Rights Watch, "US: Reassess Targeted Killings in Yemen," http://www.hrw.org/news/2013/10/21/us-reassess-targeted-killings-yemen, and Amnesty International, "USA must be held to account for drone killings in Pakistan," http://www.amnestyusa.org/news/news-item/usa-must-be-held-to-account-for-drone-killings-in-pakistan, news releases, October 2013.
16. United Nations, "Drone attacks: UN rights experts express concern about the potential illegal use of armed drones," news release, October 25, 2013, New York, http://www.ohchr.org/EN/NewsEvents/Pages/DisplayNews.aspx?NewsID=13905&LangID=E.
17. Human Rights Watch, "A Wedding That Became a Funeral, US Drone Attack on Marriage Procession in Yemen," February 2014.
18. Greg Miller, "Under Obama, a drone network," *The Washington Post*, Dec. 28, 2011.
19. Medea Benjamin, *Drone Warfare, Killing by Remote Control*, updated edition, Verso, London, 2013, p. 7. First published by OR Books, New York and London, 2012.
20. Jonathan Masters, "Backgrounder: Targeted Killings," Council on Foreign Relations, May 23, 2013.
21. Woodrow Wilson Center for International Scholars, April 30, 2012, http://www.wilsoncenter.org/event/the-efficacy-and-ethics-us-counterterrorism-strategy.
22. Rand Paul, "Sen. Paul Reaches Victory Through Filibuster," news release, March 7, 2013, http://www.paul.senate.gov/?p=press_release&id=735.
23. The White House Office of the Press Secretary, "Remarks by the President at the National Defense University," transcript, May 23, 2013, Washington, DC, http://www.whitehouse.gov/the-press-office/2013/05/23/remarks-president-national-defense-university.
24. Ryan J. Reilly, "Medea Benjamin Surprised Obama Drones Speech Protest Worked," *The Huffington Post*, May 23, 2013, http://www.huffingtonpost.com/2013/05/23/medea-benjamin-obama_n_3328321.html.
25. Charles Ellis, "Update: Former U.S. Attorney General Ramsey Clark testifies at DeWitt trial of Hancock Field drone protesters," *The Post Standard*, Syracuse.com, November 3, 2011, http://www.syracuse.com/news/index.ssf/2011/11/former_us_attorney_general_ram.html.
26. Medea Benjamin, *Drone Warfare, Killing by Remote Control*, pp 165–189.
27. Melinda Henneberger, "From Pakistan, family comes to tell of drone strike's toll,"

The Washington Post, October 29, 2013, http://www.washingtonpost.com/world/national-security/from-pakistan-family-comes-to-tell-of-drone-strikes-toll/2013/10/29/453fb990-4025-11e3-a751-f032898f2dbc_story.html.

28. Patrick Anderson, "'Unmanned': Timely tale about the human costs of using drones in war," *The Washington Post*, August 6, 2014.
29. Ryan Parker, "Drone hunting measure rejected Tuesday in Deer Trail," *The Denver Post*, April 1, 2014, http://www.denverpost.com/news/ci_25469117/drone-hunting-measure-being-voted-tuesday.
30. Aerospace Industries Association, "Poll demonstrates popular support for civil unmanned aircraft systems," news release, June 19, 2013, http://www.aia-aerospace.org/assets/FINAL_Christian_Sc._Monitor_Poll_Powerpoint.pdf.
31. Pew Research Center, "U.S. Views of Technology and the Future, Science in the next 50 years," April 17, 2014, http://www.pewinternet.org/2014/04/17/us-views-of-technology-and-the-future/.
32. Stephen Trimble, "Hijacked by Code Pink," FlightGlobal, August 11, 2011, http://www.flightglobal.com/blogs/the-dewline/2011/08/hijacked-by-code-pink/.
33. Medea Benjamin, "At Drone Convention, Zero Tolerance for Peace," *Pink Tank*, August 9, 2012, http://codepink.org/blog/2012/08/at-drone-convention-zero-tolerance-for-peace/.
34. Code Pink, "Code Pink Protester Disrupts Drone Lobby Summit, Says It's Time to Ground The Drones," news release, August 13, 2013.

Chapter 3

1. G.L. Stephens et al., "The Department of Energy's Atmospheric Radiation Measurement Unmanned Aerospace Vehicle Program," *Bulletin of the American Meteorological Society*, Vol. 81, No. 12, December 2000.
2. US Department of Energy, "Federal Aviation Regulations Parts Validation," prepared by US Department of Transportation Volpe National Transportation Systems Center and LB&M Associates, December 1994.
3. National Aeronautics and Space Administration, "NASA Armstrong Fact Sheet: Environment Research Aircraft and Sensor Technology (ERAST)," February 28, 2014, http://www.nasa.gov/centers/armstrong/news/FactSheets/FS-020-DFRC.html#.VB2i8ufB0t4.
4. National Aeronautics and Space Administration, "NASA-Industry Alliance Initiates UAV National Airspace Access Project," news release, May 20, 2004, http://www.nasa.gov/centers/dryden/news/NewsReleases/2004/04-27.html#.VCcBlefB0t4.
5. General Atomics Aeronautical Systems, "General Atomics Aeronautical Systems' Altair Receives FAA's First Commercial UAS Airworthiness Certificate," news release, September 28, 2005.
6. Federal Aviation Administration, AFS-400 UAS Policy 05-01, "Unmanned Aircraft Systems Operations in the US National Airspace System-Interim Operational Approval Guidance," September 16, 2005.
7. National Oceanic and Atmospheric Administration, "UAS Past Projects," http://uas.noaa.gov/demonstrations.html.
8. Public Employees for Environmental Responsibility, "Forest Service Drones Stuck in Hangar for 7 Years May Take Off," news release, December 3, 2013, http://www.peer.org/news/news-releases/2013/12/03/forest-service-drones-stuck-in-

hangar-for-7-years-may-take-off/.
9. US Department of Defense, "Unmanned Aircraft Systems Roadmap 2005–2030," August 2005, Washington, DC.
10. Ibid.
11. US Coast Guard, "Acquisition Update: Coast Guard Nets Historic Drug Interdiction During Most Recent Phase of Unmanned Aircraft System Technology Demonstrations," news release, July 2, 2013, http://www.uscg.mil/acquisition/newsroom/updates/uas070213.asp.
12. General Atomics Aeronautical Systems, "General Atomics Aeronautical Systems Awarded DHS/CBP UAS Contract," news release, September 1, 2005, http://media.ga.com/2005/09/01/general-atomics-aeronautical-systems-awarded-dhscbp-uas-contract/.
13US National Transportation Safety Board, Accident CHI06MA121, Oct. 31, 2007, http://www.ntsb.gov/aviationquery/brief.aspx?ev_id=20060509X00531&key=1.
14. US National Transportation Safety Board, accident reports DCA09FA009, DCA09FA028, and DCA13CA088.
15. Brian Bennett, "Police employ Predator drone spy planes on home front," *Los Angeles Times*, December 10, 2011, http://articles.latimes.com/2011/dec/10/nation/la-na-drone-arrest-20111211.
16. US Department of Homeland Security, "Privacy Impact Assessment for the Aircraft Systems," September 9, 2013, Washington, DC.
17. CBS News, "FBI director acknowledges domestic drone use," June 19, 2013, http://www.cbsnews.com/news/fbi-director-acknowledges-domestic-drone-use/.
18. Federal Bureau of Investigation, Stephen D. Kelly letter to Sen. Rand Paul, July 19, 2013, at http://www.paul.senate.gov/?p=press_release&id=905.
19. Lucy Stein, Citizens for Responsibility and Ethics in Washington, "CREW Forces FBI to Release Documents on Domestic Drone Use," news release, December 19, 2013, .
20. Citizens for Responsibility and Ethics in Washington, "Responsive document-CREW: Regarding FBI's Use of Drones in the United States: 12/2/2013-12186440-White Paper on UAS," http://www.scribd.com/doc/188717981/Responsive-document-CREW-Regarding-FBI-s-Use-of-Drones-in-the-United-States-12-2-2013-1218644-0-White-Paper-on-UAS.
21. US Department of Justice, Office of the Inspector General, "Interim Report on the Department of Justice's Use and Support of Unmanned Aircraft Systems," September 2013, Washington, DC.
22. US Department of Justice, National Institute of Justice, "Technical Bulletin: Law Enforcement Use of UAS," October 10, 2007, https://www.justnet.org/pdf/UAS-Technical-Bulletin101007.pdf.
23. MuckRock, "FOI Request, Gaston County, NC Police Drone Documents," https://www.muckrock.com/foi/gaston-county-354/gaston-county-nc-police-drone-documents-2353/#685149-responsive-documents.
24. WCNC Charlotte, "Gaston Police Department's dormant drone draws notice," March 10, 2013, http://www.wcnc.com/story/news/local/2014/07/02/10979942/.
25. US Department of Justice, National Institute of Justice, National Law Enforcement and Corrections Technology Center, "TechBeat," summer 2011, https://www.justnet.org/InteractiveTechBeat/summer_2011/DHSSAVERProgram.pdf.

26. US Federal Emergency Management Agency, "DHS Authorized Equipment List," March 7, 2012, http://www.fema.gov/media-library-data/20130726-1825-25045-7138/fema_preparedness_grants_authorized_equipment_list.pdf.
27. Bill Carey, "Companies Demonstrate Small UAS for DHS Program," *AINonline*, March 29, 2013, http://www.AINonline.com/aviation-news/2013-03-29/companies-demonstrate-small-uas-dhs-program.
28. Kirk Kloeppel, "Possible UAS Future for the Department of Homeland Security," PowerPoint presentation to the 2014 AUVSI Unmanned Systems Conference, May 12, 2014.
29. US Department of Justice, Bureau of Justice Statistics, "Special Report: Aviation Units in Large Law Enforcement Agencies, 2007," July 2009, http://www.bjs.gov/content/pub/pdf/aullea07.pdf.
30. Mesa County, Colo., Sheriff's Office, "Unmanned Aerial System Team," http://sheriff.mesacounty.us/uav/.

.

Chapter 4

1. Bill Carey, "Pressure Mounts on U.S. Congress To Fund FAA," *AINonline*, September 12, 2011, http://www.AINonline.com/aviation-news/ain-air-transport-perspective/2011-09-12/pressure-mounts-us-congress-fund-faa.
2. Bill Carey, "Rockwell Collins CEO Urges New Approach to NextGen," *AINonline*, June 6, 2012, http://www.AINonline.com/aviation-news/blogs/ain-blog-rockwell-collins-ceo-urges-new-approach-nextgen.
3. Federal Aviation Administration, "Fact Sheet," July 2011, http://www.faa.gov/about/initiatives/uas/media/uas_fact_sheet.pdf).
4. US Rep. John Mica, "Long Stalled FAA Reform Measure Signed into Law," news release, February 14, 2012, http://mica.house.gov/current-activities2/long-stalled-faa-reform-measure-signed-into-law.
5. Bart Jansen, "FAA directed to make (air)space for drones," *USA Today*, February 7, 2012, http://usatoday30.usatoday.com/news/nation/story/2012-02-06/unmanned-drones-share-faa-airspace/52994752/1.
6. Lee Fang, "Exclusive: PowerPoint Shows Drone Industry's Lobbying Plan to Expand Over Domestic, Law Enforcement Markets," Republic Report, February 15, 2012, http://www.republicreport.org/2012/drone-powerpoint-lobby-plan.
7. Gary Martin and Viveca Novak, "Drones: Despite Problems, A Push to Expand Domestic Use," Center for Responsive Politics OpenSecrets Blog, November 27, 2012, www.opensecrets.org.
8. Government Accountability Office, "Unmanned Aircraft Systems, Measuring Progress and Addressing Potential Privacy Concerns Would Facilitate Integration into the National Airspace System," Washington, DC, September 2012, http://www.gao.gov/assets/650/648348.pdf.
9. Federal Aviation Administration, "Integration of Civil Unmanned Aircraft Systems in the National Airspace System Roadmap," Washington, DC, November 7, 2013, http://www.faa.gov/about/initiatives/uas/media/uas_roadmap_2013.pdf.
10. US Department of Transportation, "Unmanned Aircraft Systems Comprehensive Plan: A Report on the Nation's UAS Path Forward," Washington, DC, November 6, 2013. http://www.faa.gov/about/office_org/headquarters_offices/agi/reports/media/UAS_Comprehensive_Plan.pdf.

11. Bill Carey, "FAA Designates Six Test Ranges for Unmanned Aircraft," *AINonline*, December 30, 2013, http://www.AINonline.com/aviation-news/2013-12-30/faa-designates-six-test-ranges-unmanned-aircraft.
12. U.S. Department of Transportation Office of the Inspector General, "Top Management Challenges for Fiscal Year 2014," December 16, 2013, Washington, DC, http://www.oig.dot.gov/library-item/6262.
13. Government Accountability Office, "FAA Reauthorization Act: Progress and Challenges Implementing Various Provision of the 2012 Act," Washington, DC, February 5, 2014, http://www.gao.gov/assets/670/660683.pdf.
13. House Appropriations Committee, House Report 113-464 to accompany H.R. 4745, May 27, 2014.
14. Senate Appropriations Committee, Senate Report 113-182 to accompany S.2438, June 5, 2014.
15. US Department of Transportation Office of the Inspector General, "FAA faces significant barriers to safely integrate unmanned aircraft systems into the National Airspace System," audit report, June 26, 2014, Washington, DC.
16. US Department of Transportation, Office of the Inspector General, "FAA's Progress and Challenges in Integrating Unmanned Aircraft Systems into the National Airspace System," testimony of Assistant Inspector General for Aviation Matthew Hampton before the House Subcommittee on Aviation, December 10, 2014.

.

Chapter 5

1. Sen. John Cornyn, "Cornyn Announces FAA Approval of Predator Flights Along Entire Texas-Mexico Border," news release, June 23, 2010.
2. Federal Aviation Administration, "Safety Must Come First," Administrator J. Randolph Babbitt speech, November 18, 2009, http://www.faa.gov/news/speeches/news_story.cfm?newsId=10964.
3. Federal Aviation Administration, "More Critical Aviation Projects Come to Halt Due to Congressional Inaction," news release, July 26, 2011, Washington, DC.
4. Federal Aviation Administration, "Statement from Randy Babbitt," news release, December 6, 2011, Washington, DC.
5. Humberto Sanchez, "Jim DeMint: No More Appointees Until After Elections," *Roll Call*, August 1, 2012, http://www.rollcall.com/news/jim_demint_no_more_appointees_until_after_elections-216663-1.html.
6. US Senate Committee on Commerce, Science and Transportation, Nomination Hearing, June 21, 2012, http://www.commerce.senate.gov/public/.
7. Econsult Corporation, "Economic Impacts of FAA Budget Sequestration on U.S. Economy," submitted to the Aerospace Industries Association, August 2012.
8. Federal Aviation Administration, "Safely Integrating Unmanned Aircraft Systems," Michael Huerta speech to AUVSI Unmanned Systems Conference, August 7, 2012, Las Vegas.
9. Timothy Adelman, "Under New Leadership, FAA's Unmanned Aircraft Systems Integration Office Meets Its Deadline," *AUVSI Unmanned Systems News*, May 14, 2012.
10. Bill Hutchinson, "Drone captures scene at East Harlem explosion that flattened two buildings," *New York Daily News*, March 13, 2014, http://www.nydailynews.com/new-york/uptown/drone-captures-e-harlem-explosion-scene-video-

article-1.1719988.

11. US Department of Transportation Inspector General, "FAA's Implementation of the FAA Modernization and Reform Act of 2012 Remains Incomplete," Calvin L. Scovel III testimony before the House Aviation Subcommittee, February 5, 2014, Washington, DC.
12. US Government Accountability Office, "FAA Reauthorization Act, Progress and Challenges Implementing Various Provisions of the 2012 Act," Gerald L. Dillingham testimony before the House Aviation Subcommittee, February 5, 2014, Washington, DC.

Chapter 6

1. Box Office Mojo, "The Aviator," http://www.boxofficemojo.com/movies/?id=aviator.htm.
2. Federal Aviation Administration, *Advisory Circular 91-57*, June 9, 1981, Washington, DC.
3. Federal Aviation Administration, Docket No. FAA-2006-25714, "Unmanned Aircraft Operations in the National Airspace System," notice of policy, February 6, 2007, Washington, DC.
4. Federal Aviation Administration, Order 1110.150, "Small Unmanned Aircraft System Aviation Rulemaking Committee," April 10, 2008, Washington, DC.
5. Federal Aviation Administration, "Small Unmanned Aircraft System Aviation Rulemaking Committee, Comprehensive Set of Recommendations for sUAS Regulatory Development," April 1, 2009, http://www.faa.gov/regulations_policies/rulemaking/committees/documents/media/SUAS.ARC.RR.20090401.pdf.
6. Exec. Order No. 12866, 58 Fed. Reg. 190 (Oct. 4, 1993).
7. ASTM International, "Major Facets of Small Unmanned Aircraft Systems Operations Covered in Seven New ASTM Standards," news release, April 1, 2014, http://www.astmnewsroom.org/default.aspx?pageid=3408.
8. Patrick Egan, "UAS ARC 2.0," *sUAS News*, Nov. 27, 2011, http://www.suasnews.com/2011/11/10245/uas-arc-2-0/.
9. Federal Aviation Administration, "Aviation Rulemaking Committee Charter," June 11, 2014, http://www.faa.gov/regulations_policies/rulemaking/committees/documents/media/UAS.ARC.Cht.Ext.20140611.pdf.
10. Huerta v. Pirker, NTSB Docket CP-217, Respondent's Motion to Dismiss, September 27, 2013.
11. Huerta v. Pirker, NTSB Docket CP-217, Decisional Order, March 6, 2013.
12. Jason Koebler, "Here's Why Commercial Drones Just Got Legalized," *Motherboard*, March 7, 2014, http://motherboard.vice.com/read/commercial-drones-just-got-legalized-heres-why.
13. Ashley Downs, "Commercial Drone Ban Lifted. Kindle by Predator Greenlit," *News Ledge*, March 7, 2014, http://www.newsledge.com/commercial-drone-ban-lifted-kindle-predator-greenlit-5395.
14. Huerta v. Pirker, NTSB Order No. EA-5730, November 18, 2014.
15. Craig Whitlock, "FAA staff overruled on drone safety fears," *The Washington Post*, December 22, 2014.
16. Federal Aviation Administration, "Petition of UAS America Fund, LLC, to Adopt 14 C.F.R. Part 107 to Implement Operational Requirements for Micro Unmanned

Aircraft Systems," December 18, 2014.

Chapter 7

1. The Wire, "Was the Drone that Scared Feinstein at Her House This Tiny Pink Helicopter?", January 15, 2014, http://www.thewire.com/politics/2014/01/was-drone-feinstein-encountered-her-house-tiny-pink-helicopter/357061/.
2. Peter Finn, "Domestic use of aerial drones by law enforcement likely to prompt privacy debate," *The Washington Post*, January 23, 2011, http://www.washingtonpost.com/wp-dyn/content/article/2011/01/22/AR2011012204111.html.
3. Electronic Frontier Foundation, "FAA Releases Lists of Drone Certificates—Many Questions Left Unanswered," news release, April 19, 2012, https://www.eff.org/deeplinks/2012/04/faa-releases-its-list-drone-certificates-leaves-many-questions-unanswered.
4. Department of Homeland Security, Office of Inspector General, "CBP's Use of Unmanned Aircraft Systems in the Nation's Border Security," May 30, 2012, Washington, DC.
5. Electronic Frontier Foundation v. Department of Homeland Security, "Complaint for Injunctive Relief for Violation of the Freedom of Information Act," US District Court for the Northern District of California, October 30, 2012, https://www.eff.org/files/filenode/CBPComplaint.pdf.
6. Electronic Frontier Foundation, "Customs & Border Protection Logged Eight-Fold Increase in Surveillance for Other Agencies," news release, July 3, 2013, https://www.eff.org/deeplinks/2013/07/customs-border-protection-significantly-increases-drone-surveillance-other.
7. Electronic Frontier Foundation, "Customs & Border Protection Loaned Drones to Other Agencies 700 Times in Three Years According to 'Newly Discovered' Records," news release, January 14, 2014.
8. US Department of Homeland Security, "Concept of Operations for CBP's Predatory Aircraft System," June 29, 2010, via Electronic Frontier Foundation Freedom of Information Act lawsuit, October 30, 2012.
9. Association for Unmanned Vehicle Systems International, "Unmanned Aircraft System Operations Industry Code of Conduct," July 2, 2012, http://www.auvsi.org/conduct.
10. International Association of Chiefs of Police Aviation Committee, "Recommended Guidelines for the use of Unmanned Aircraft," August 2012, http://www.theiacp.org/portals/0/pdfs/IACP_UAGuidelines.pdf.
11. American Civil Liberties Union, "The Future of Unmanned Aviation in the U.S. Economy: Safety and Privacy Considerations," statement of Christopher Calabrese, legislative counsel, before the Senate Committee on Commerce, Science and Transportation," January 15, 2014.
12. M. Ryan Calo, "The Drone as Privacy Catalyst," 64 STAN. L. REV. Online 29 (2011).
13. Electronic Privacy Information Center, "Petition to the FAA, Drones and Privacy," February 24, 2012, http://epic.org/privacy/drones/FAA-553e-Petition-03-08-12.pdf.
14. California v. Ciraolo, 476 U.S. 207 (1986).
15. Dow Chemical Co. v. U.S., 476 U.S. 227 (1986).

16. Florida v. Riley 488 U.S. 445 (1989).
17. Kyllo v. U.S., 533 U.S. 27 (2001).
18. U.S. v. Jones, 132 S.Ct. 945 (2012).
19. Noel Sharkey and Sarah Knuckey, "Occupy Wall Street's 'occucopter' — who's watching whom?", *The Guardian*, December 21, 2011, http://www.theguardian.com/commentisfree/cifamerica/2011/dec/21/occupy-wall-street-occucopter-tim-pool.
20. Holland & Knight LLP, "Brief of News Media Amici in Support of Respondent Raphael Pirker," National Transportation Safety Board, Docket No. CP-217, May 6, 2014.
21. The Rutherford Institute, "Relying on Rutherford Institute Model Resolution, Charlottesville Becomes First U.S. City to Limit Police Drones; TRI Calls on Rest of Country to Follow Suit," news release, February 5, 2013.
22. Bill Carey, "Group: States Should Require Warrants for UAS Surveillance," *Aviation International News*, August 13, 2013, http://www.AINonline.com/aviation-news/2013-08-13/group-states-should-require-warrants-uas-surveillance.
23. Allie Bohm, "The First State Laws on Drones," April 15, 2013, American Civil Liberties Union, https://www.aclu.org/blog/technology-and-liberty-national-security/first-state-laws-drones.
24. Allie Bohm, "Status of 2014 Domestic Drone Legislation in the States," American Civil Liberties Union, April 22, 2014, https://www.aclu.org/blog/technology-and-liberty/status-2014-domestic-drone-legislation-states.
25. Federal Aviation Administration, "Unmanned Aircraft Systems Test Site Selection," FAA Contract Opportunities special notice, February 14, 2013, https://faaco.faa.gov/index.cfm/announcement/view/13143.
26. Federal Aviation Administration, "FAA Online Listening Session," transcript, April 3, 2013, http://www.faa.gov/about/initiatives/uas/.
27. Federal Aviation Administration, Docket No.: FAA-2013-0061, "Notice of availability of final privacy requirements for unmanned aircraft system test site program; response to comments," November 7, 2013, http://www.faa.gov/about/initiatives/uas/media/UAS_privacy_requirements.pdf.
28. American Civil Liberties Union, "The Future of Unmanned Aviation in the U.S. Economy: Safety and Privacy Considerations," statement of Christopher Calabrese, legislative counsel, before the Senate Committee on Commerce, Science and Transportation," January 15, 2014.
29. Gautam Hans, "Domestic Drones are Coming; Will Privacy be Factored In?", Center for Democracy and Technology, November 22, 2013, https://cdt.org/blog/domestic-drones-are-coming-will-privacy-be-factored-in/.
30. US Congress, "Explanatory statement submitted by Mr. Rogers of Kentucky, chairman of the House Committee on Appropriations, regarding the House amendment to the Senate amendment on H.R. 3547, Consolidated Appropriations Act, 2014," Congressional Record, Vol. 160, No. 9, January 15, 2014.
31. Erin Mershon and Kevin Robillard, "President Barack Obama to issue executive order on drone privacy," *Politico*, July 23, 2014, http://www.politico.com/story/2014/07/executive-order-drone-privacy-barack-obama-109303.html.

Chapter 8

1. Federal Aviation Administration, "FAA Aerospace Forecast, Fiscal Years 2014-2034," March 2014, Washington, D.C., pg. 60.
2. Paul Rinaldi, National Air Traffic Controllers Association, "A Review of FAA's Efforts to Reduce Costs and Ensure Safety and Efficiency Through Realignment and Consolidation," testimony before the House Aviation Subcommittee, May 31, 2012.
3. Jack L. Shagena, Jr., and James E. Dalmas, *Flying Safer Skies With Electronic Eyes* (Shagena Publishing, 2012) pg. 11.
4. Bill Carey, Introduction to *Flying Safer Skies With Electronic Eyes*, by Jack L. Shagena Jr., and James E. Dalmas (Shagena Publishing 2012).
5. Matthew T. DeGarmo, "Issues Concerning Integration of Unmanned Aerial Vehicles in Civil Airspace," The MITRE Corporation, McLean, VA, November 2004.
6. Bill Carey, "Small UAS Rule Will Begin Phased Entry of Unmanned Aircraft," *Aviation International News*, October 2011, http://www.AINonline.com/aviation-news/aviation-international-news/2011-10-04/small-uas-rule-will-begin-phased-entry-unmanned-aircraft.
7. Federal Aviation Administration, Aeronautical Surveillance Panel, "ACAS Xu, Ensuring Collision Avoidance Interoperability," Washington, DC, September 2012.
8. University of North Dakota, "UND, research partners NASA and The MITRE Corp. successfully test new UAS technology," news release, September 2012, http://und.edu/news/2012/09/und-nasa-test-uas-tech.cfm.
9. Bill Carey, "Unmanned Aircraft Will Be Introduced to Airspace Safely, Slowly," *Aviation International News*, September 2012, http://www.AINonline.com/aviation-news/aviation-international-news/2012-09-02/unmanned-aircraft-will-be-introduced-airspace-safely-slowly.
10. R³ Engineering, "R³ Engineering announces the first ADS-B fully autonomous collision avoidance demonstration for Unmanned Aircraft," news release, October 24, 2012.
11. R³ Engineering, "Unmanned Aircraft Systems—Successful Navy Funded, ADS-B based Autonomous, Sense & Avoid with Pilot Override Demonstrated," news release, April 24, 2013.
12. Sierra Nevada Corporation, "Sense and Avoid Radar Sensor Successfully Completes First Flight Demonstration," news release, November 15, 2011, http://www.sncorp.com/press_more_info.php?id=468.
13. Bill Carey, "U.S. Navy Advances Triton Without 'Sense-and-Avoid' Settled," *AINonline*, April 11, 2014, http://www.AINonline.com/aviation-news/2014-04-08/us-navy-advances-triton-without-sense-and-avoid-settled.
14. FedBizOpps.Gov, "Request for Information, Common Airborne Sense and Avoid Sources Sought and Request for Capability Statements," Solicitation FA862014R4000, September 20, 2013.
15. US Department of Defense, Instruction Number 4540.1, "Use of International Airspace by U.S. Military Aircraft and for Missile/Projectile Firings," March 28, 2007, http://www.dtic.mil/whs/directives/corres/pdf/454001p.pdf.
16. General Atomics Aeronautical Systems, "GA-ASI Successfully Tests SAA System Components Aboard Predator B," news release, December 12, 2013, http://www.ga-asi.com/news_events/index.php?read=1&id=430.

17. US Department of Defense, "Report to Congress on Future Unmanned Aircraft Systems Training, Operations and Sustainability," April 2012.
18. Viva Kelley and Larry Herbek, "Ground-Based Sense-and-Avoid Forges the Path," AUVSI Unmanned Systems, Arlington, VA, March 2014. .
19. Thomas Spriesterbach, Kelly Bruns, Lauren Baron, and Jason Sohlke, "Unmanned Aircraft System Airspace Integration in the National Airspace Using a Ground-Based Sense and Avoid System," *Johns Hopkins APL Technical Digest*, Vol. 32, No. 3 (2013), http://www.jhuapl.edu/techdigest/TD/td3203/32_03-spriesterbach.pdf.
20. Roger Francis, "Ground Based Sense and Avoid 101 Overview," presentation, The MITRE Corporation, 2011.
21. Frank Box, Warren Wilson, and Leo Globus, "Radio Spectrum Requirements of Control and Non-Payload Communications Links for Future Unmanned Aircraft Systems," *Air Traffic Control Quarterly*, Vol. 21, No. 1, 2013, Air Traffic Control Association Institute, Alexandria, Va.
22. Federal Aviation Administration, Notice 8900.227, "Unmanned Aircraft Systems (UAS) Operational Approval," July 30, 2013, Washington, DC.
23. Amy Butler, "No Scouting for Fire Scout," *Aviation Week & Space Technology*, Ares defense technology blog, August 25, 2010.
24. NASA Aviation Safety Reporting System Database Report Set, "Unmanned Aerial Vehicle Reports," August 11, 2013, Ames Research Center, Moffett Field, California.
25. Federal Aviation Administration, "Integration of Unmanned Aircraft Systems into the National Airspace System Concept of Operations V2.0," September 28, 2012, Washington, DC.
26. The University of Texas at Austin, "Todd Humphreys' Research Team Demonstrates First Successful GPS Spoofing of UAV," news release, July 2012, http://www.ae.utexas.edu/news/features/todd-humphreys-research-team-demonstrates-first-successful-gps-spoofing-of-uav.
27. Adaptive Flight, "Adaptive Flight Works with University of Texas for GPS Spoofing Tests—Announces GPS-Denied Software," news release, July 12, 2012, http://www.adaptiveflight.com/adaptive-flight-works-with-university-of-texas-for-gps-spoofing-tests-announces-gps-denied-software.
28. Department of Transportation Inspector General, "FAA Faces Significant Barriers to Safely Integrate Unmanned Aircraft Systems into the National Airspace System," audit report, June 26, 2014, Washington, DC.
29. Federal Aviation Administration, "Multi-UAS Operational Assessment: Class D Airspace," July 2012.
30. RTCA Paper No. 08-072/SC203-029, "Summary of the Twelfth Meeting, Special Committee 203," meeting minutes, March 19, 2008, Washington, DC.
31. RTCA Paper No. 026-13/SC203-061, "Summary of the Twenty Third Meeting, Special Committee 203, meeting minutes, March 11, 2013.
32. Bill Carey, "New RTCA Committee Seeks to Expedite UAS Standards," *AINonline*, April 5, 2013, http://www.AINonline.com/aviation-news/2013-04-05/new-rtca-committee-seeks-expedite-uas-standards.
33. RTCA Paper No. 096-13/PMC-1083, "Program Management Committee Meeting Summary, March 20, 2013," May 20, 2013, Washington, DC, http://www.rtca.

org/Files/Committee%20Meeting%20Summaries/PMC%20Mar%202013%20 Summary.pdf.

34. Bill Carey, "New RTCA Committee Seeks to Expedite UAS Standards," *AINonline*, April 5, 2013, http://www.AINonline.com/aviation-news/2013-04-05/new-rtca-committee-seeks-expedite-uas-standards.

.

Chapter 9

1. Federal Aviation Administration, "Selection of six unmanned aircraft systems test sites in accordance with FAA Modernization and Reform Act of 2012, PL-112-95," Order, December 30, 2013.
2. Government Accountability Office, "Unmanned Aircraft Systems, Measuring Progress and Addressing Potential Privacy Concerns Would Facilitate Integration into the National Airspace System," September 2012, Washington, DC.
3. US Energy Information Administration, "Five states and the Gulf of Mexico produce more than 80% of U.S. crude oil," news release, March 31, 2014, http://www.eia.gov/todayinenergy/detail.cfm?id=15631#.
4. Lone Star UAS Center of Excellence & Innovation website, http://lsuasc.tamucc.edu/outreach.html.
5. Bill Carey, "FAA's 'Vision' Is To Delegate Authority to UAS Test Ranges," *AINonline*, May 5, 2014, http://www.AINonline.com/aviation-news/2014-05-05/faas-vision-delegate-authority-uas-test-ranges.

.

Chapter 10

1. CBS News, "Amazon unveils futuristic plan: Delivery by drone," *60 Minutes*, December 1, 2013, http://www.cbsnews.com/videos/amazon-unveils-futuristic-plan-delivery-by-drone/.
2. Mary Cummings, "Drones on and on and on, US needs to join the aviation revolution," *The Boston Globe*, December 8, 2013, http://www.bostonglobe.com/opinion/2013/12/08/drones-and-and/MgLXPHtr0nvWqwkltSORYM/story.html.
3. ARD-aktuell/tagesschau.de, "Deutsche Post DHL tests drone by 'Paketcopter' to the other bank," December 9, 2013, http://www.tagesschau.de/wirtschaft/paketcopter102.html.
4. Konstantin Kakaes, "Amazon Prime Drone Delivery? It's Hot Air," *Slate*, December 2, 2013, http://www.slate.com/blogs/future_tense/2013/12/02/amazon_prime_drone_delivery_service_jeff_bezos_plan_is_full_of_hot_air.html.
5. Federal Aviation Administration, "FAA Aerospace Forecast Fiscal Years 2013 2033," March 19, 2013, Washington, DC.
6. Internet.org, "Inventing the Future of Connectivity, YouTube video, March 27, 2014, https://www.youtube.com/watch?v=pxX6r-xDgG4&feature=youtu.be&noredirect=1.
7. The companies awarded Section 333 exemptions by the FAA in December 2014 were: Trimble Navigation, of Sunnyvale, California, for the UX5; Clayco, of St. Louis, for the Skycatch quadcopter; VDOS Global of Corvallis, Oregon, for the Aeryon SkyRanger; and Woolpert, of Annapolis, Maryland, for the Altavian Nova Block III airplane.
8. Krithika Krishnamurthy, "India to be launch pad for Amazon's plan to deliver packages using drones; deliveries may start by Diwali," *The Economic Times*,

August 20, 2014, http://articles.economictimes.indiatimes.com/2014-08-20/news/53028827_1_prime-air-drones-outdoors-amazon.

9. Federal Aviation Administration, Docket FAA-2014-0474, "Amazon.com-Additional Information," December 8, 2014, at regulations.gov, http://www.regulations.gov/#!docketDetail;D=FAA-2014-0474.
10. Bill Carey, "AeroVironment, Lockheed Martin Advance Global Observer UAS," *AINonline*, February 9, 2014, http://www.AINonline.com/aviation-news/2014-02-09/aerovironment-lockheed-martin-advance-global-observer-uas.
11. CBS News, "Drones Over America," *60 Minutes*, March 16, 2014, http://www.cbsnews.com/news/drones-over-america-60-minutes/.
12. Teal Group, "2014 World Unmanned Aerial Vehicles Systems," Executive Overview, July 2014, Fairfax, VA.
13. Darryl Jenkins and Bijan Vasigh, "The Economic Impact of Unmanned Aircraft Systems Integration in the United States," Association for Unmanned Vehicle Systems International, March 2013, Arlington, VA.
14. Measure, "American Farm Bureau Federation and Measure Produce First Ever Report and Calculator on Value of Drones in Agriculture," news release, July 21, 2015, http://measure32.com/measure-afbf-roicalculator-drones-precisionag/.
15. Yamaha Motor Co. Ltd., "Yamaha Motor Monthly Newsletter," November 18, 2013, Issue No. 11.
16. University of California Cooperative Extension, "Vine Lines," August 2013, Fresno, Calif.
17. Henio Arcangeli, testimony before the Senate Transportation Committee hearing, "The Future of Unmanned Aviation in the U.S. Economy: Safety and Privacy Considerations," January 15, 2014.
18. Benjamin Litvinas and Nathan Dickerson, "Analyzing the Business Case and Economic Viability of Unmanned Aircraft within the Nation's Airspace," American Institute of Aeronautics and Astronautics, August 14, 2013.
19. Mickey Osterreicher, "Charting the Course for Use of Small Unmanned Aerial Systems in Newsgathering," National Press Photographers Association, May 2014, Buffalo, NY.
20. AUVSI News, "CNN Deploys a Drone For Historic Selma march Coverage," Association for Unmanned Vehicle Systems International, March 5, 2015, http://www.auvsi.org/Mississippi/blogs/auvsi-news/2015/03/05/cnnheli.
21. Craig Whitlock, "When Drones Fall From The Sky," *The Washington Post*, June 20, 2014, http://www.washingtonpost.com/sf/investigative/2014/06/20/when-drones-fall-from-the-sky/?hpid=z1.
22. Aerospace Industries Association, "'When Drones Fall from the Sky' Falls Short," news release, June 20, 2014.
23. Association for Unmanned Vehicle Systems International, "Washington Post Article 'When Drones Fall from the Sky' Misses the Mark," June 30, 2014.
24. Mary Cummings, testimony before the Senate Transportation Committee hearing, "The Future of Unmanned Aviation in the US Economy: Safety and Privacy Considerations," January 15, 2014.
25. Craig Whitlock, "Drone close calls with aircraft surge," *The Washington Post*, November 27, 2014.

26. Sen. Dianne Feinstein, "Feinstein Introduces Bill to Improve Safety of Consumer Drones," news release, June 18, 2015, Washington, DC, http://www.feinstein.senate.gov/public/index.cfm/press-releases?ID=8C18CEB0-4CA0-4A1D-A251-CF8E9E94C06E.
27. Matt Hamilton and Veronica Rocha, "$75,000 in rewards offered to catch operators who flew drones above fires," *The Los Angeles Times*, July 28, 2015, http://www.latimes.com/local/lanow/la-me-ln-reward-drone-arrests-20150728-story.html.
28. Sen. Ted Gaines, "Senator Gaines and Assemblyman Gatto Announce Bill to Knock Drones Out of Fire Zones," news release, July 9, 2015, http://gaines.cssrc.us/content/senator-gaines-and-assemblyman-gatto-announce-bill-knock-drones-out-fire-zones.
29. Federal Aviation Administration, "Pilot Unmanned Aircraft Reports Much Higher in 2015," news release, August 12, 2015, Washington, DC, http://www.faa.gov/news/updates/?newsId=83445.
30. Jack Nicas, "Pilots Are Spotting More Drones, Prompting Worries About Collisions," *The Wall Street Journal*, August 13, 2015, http://www.wsj.com/articles/pilots-are-spotting-more-drones-prompting-worries-about-collisions-1439498782.
31. Chris Manno, "Airliners vs. Drones: Calm Down," *JetHead* blog, https://jethead.wordpress.com/2015/08/08/airliners-vs-drones-calm-down/.
32. Karen Walker, "A drone is going to bring down an airliner: why are we waiting for that to happen?", *Air Transport World*, July 21, 2015, http://atwonline.com/blog/drone-going-bring-down-airliner-why-are-we-waiting-happen.
33. Edward Wytkind, "We Said Go-Slow on Drones. We Weren't Kidding," Transportation Trades Department, AFL-CIO, August 4, 2015, http://ttd.org/blog/we-said-go-slow-on-drones-we-werent-kidding/.
34. UAS Registration Task Force Aviation Rulemaking Committee, "Task Force Recommendations Final Report," November 21, 2015, https://www.faa.gov/uas/publications/media/RTFARCFinalReport_11-21-15.pdf .
35. Federal Aviation Administration, "It's (a) Grand! FAA Passes 1,000 UAS Section 333 Exemptions," news release, August 4, 2015, Washington, DC, http://www.faa.gov/news/updates/?newsid=83395.
36. National Aeronautics and Space Administration, "NASA Signs Lease with Planetary Ventures LLC for Use of Moffett Airfield and Restoration of Hangar One," news release, November 10, 2014, Washington, DC, http://www.nasa.gov/press/2014/november/nasa-signs-lease-with-planetary-ventures-llc-for-use-of-moffett-airfield-and/#.VeG_es7q0iQ.
37. Amazon, "Revising the Airspace Model for the Safe Integration of Small Unmanned Aircraft Systems," NASA UTM 2015 convention, July 2015, http://utm.arc.nasa.gov/docs/Amazon_Revising%20the%20Airspace%20Model%20for%20the%20Safe%20Integration%20of%20sUAS%5B6%5D.pdf.
38. Northrop Grumman, "Unmanned Systems: The Value of Innovation," CEO Wes Bush speech to AUVSI North America 2011 Forum, Washington, DC, August 17, 2011, http://www.northropgrumman.com/MediaResources/Presentations/2011/Pages/081711WesBushAtAuvsi2011Forum.aspx.
39. US Department of Commerce Office of Export Enforcement, "Dutch Company Agrees to Pay $10.5 Million Civil Penalty to Settle Commerce Department Charges Involving Illegal Exports and Reexports to Iran and Sudan," news release, June

5, 2014, http://www.bis.doc.gov/index.php/component/content/article/9-bis/carousel/687-dutch-company-agrees-to-pay-10-5-million-civil-penalty-to-settle-commerce-department-charges-involving-illegal-exports-and-reexports-to-iran-and-sudan.

.

CHAPTER 11

1. BBC News, "Fears over unmanned spy planes," July 23, 2004, http://news.bbc.co.uk/2/hi/uk_news/wales/mid_/3919219.stm.
2. West Wales Airport, "CAA approves dedicated UK airspace for unmanned aircraft for first time," news release, June 7, 2011.
3. Medea Benjamin, "Drone Warfare, Killing by Remote Control," Verso 2013, pgs. 193–194.
4. BEPJ, "Drone crashes at Parc Aberporth—the twisted shape of things to come?" September 23, 2009, http://www.bepj.org.uk/drone-crashes-at-parc-aberporth.
5. Angus Batey, "UK seen as RPAS integration leader," *AIN Farnborough Airshow News*, July 14, 2014, http://www.AINonline.com/aviation-news/farnborough-air-show/2014-07-12/british-regulator-leads-rpas.
6. Safran Sagem, "Sagem demonstrates Patroller drone's ability to operate in civilian airspace," news release, November 20, 2014, http://www.sagem.com/spip.php?article1264.
7. Lambert Dopping-Hepenstal, "Unmanned Aerial Vehicles," *Ingenia*, June 2014, http://www.ingenia.org.uk/ingenia/articles.aspx?index=903.
8. Alan Warnes, "Turkish firm shows attack helo and UAV," *AIN Farnborough Airshow News*, July 14, 2014, www.AINonline.com/aviation-news/farnborough-air-show/2014-07-13/tai-exhibits-attack-helicopter-and-anka-uav.
9. Chris Pocock, "Neuron UCAV Makes Its Maiden Flight," *AINonline*, December 3, 2012, http://www.AINonline.com/aviation-news/ain-defense-perspective/2012-12-03/neuron-ucav-makes-its-maiden-flight.
10. David Donald, "UK Finally Reveals Flight Trials of Taranis UCAS," *AINonline*, February 5, 2014, www.AINonline.com/aviation-news/ain-defense-perspective/2014-02-05/uk-finally-reveals-flight-trials-taranis-ucas.
11. David Donald, "New players in large-UAV game," *AIN Farnborough Airshow News*, July 15, 2014, www.AINonline.com/aviation-news/farnborough-air-show/2014-07-15/uavs-new-players-emerging.
12. UVS International, "Remotely Piloted Aircraft Systems Integration into EU Airspace," European RPAS Policy & Regulatory Forum abstracts, June 2014, http://rpas-2014.org/phocadownloadpap/07_all_docs/RPAS-2014_Conference-Program_Long_140520_HR.pdf.
13. Angus Batey, "UK seen as RPAS integration leader," *AIN Farnborough Airshow News*, July 14, 2014, www.AINonline.com/aviation-news/farnborough-air-show/2014-07-12/british-regulator-leads-rpas.
14. Australia Parliament, House of Representatives Standing Committee on Social Policy and Legal Affairs, "Eyes in the sky, Inquiry into drones and the regulation of air safety and privacy," July 2014, Canberra.
15. Australian Certified UAV Operators Inc., "Australia now has more than 100 certified unmanned aircraft operators," news release, May 18, 2014.

16. Australia Civil Aviation Safety Authority, Notice of Proposed Rulemaking 1309OS, "Remotely Piloted Aircraft Systems," May 2014.
17. Australian Certified UAV Operators Inc., "'Eyes In the Sky' Report Is Timely But Confuses Impact of RPAS Deregulation With Opportunities for Enhanced Privacy Structures," news release, July 14, 2014.
18. European RPAS Steering Group, "Roadmap for the integration of civil Remotely-Piloted Aircraft Systems into the European Aviation System," June 2013.
19. European Commission, "European Commission calls for tough standards to regulate civil drones," news release, April 8, 2014, Brussels, Belgium, http://ec.europa.eu/ireland/press_office/news_of_the_day/eu-calls-for-tough-standards-on-civil-drones_en.htm.
20. Siim Kallas, "Statement to the media," news release, April 8, 2014, Brussels, Belgium, http://europa.eu/rapid/press-release_STATEMENT-14-110_en.htm.

ABOUT THE AUTHOR

Bill Carey began his professional career as a newspaper reporter and for the past decade has worked in the aviation trade press. He has written extensively about unmanned aerial vehicles, initially as editor-in-chief of *Avionics* magazine, and more recently as senior editor with *Aviation International News*, based in Washington, DC.